CROSSFEED

CROSSFEED

Professional Solutions
to Common Flying Problems

by DAN MANNINGHAM

Jason Aronson, Inc.
New York & London

Much of the material in this book first appeared in *Business and Commercial Aviation* Magazine. Technical illustrations were prepared by the *Business and Commercial Aviation* art staff.

Library of Congress Cataloging Publication Data

Manningham, Dan.
  Crossfeed.

  1. Aeronautics—Popular works. I. Title.
TL546.7.M32 629.13 82-6698
ISBN 0-87668-602-1   AACR2

Manufactured in the United States of America.

I would like to thank Archie,
who said all this would lead to a book.

Contents

CONTENTS

CONTENTS

Introduction

It is barely morning. The eastern sky glows orange, but the sun has not yet surfaced as an F-4 is catapulted aloft from an aircraft carrier somewhere in the Pacific. One hour later and halfway to Barbers Point Naval Air Station, this crew is in trouble. Fuel will not transfer properly, and they must nurse every available pound from the functioning tanks in order to avoid ditching. In short, they need to know and employ the most fuel-conservative flight profile from their present position to the nearest airport. When they land—safely, as it turns out—their story will illustrate a classic challenge of aviation.

It is evening in London. The crew of this Gulfstream II is employed by one of America's major corporations and has been asked to conduct a flight over the North Pole to Fairbanks, Alaska. Because the G-II's range is marginal at best for this stage length, the flight has been planned with extraordinary emphasis on cruise control techniques. When they return home, these pilots will have learned priceless lessons in the art of maximizing their airplane's range capability.

It is anytime, at Chicago's O'Hare Airport. Holding stacks at FARMM, PLANT and PAPPI intersections are all full. Total delay time is over an hour for inbound flights as an airline crew makes their first turn in the holding pattern. Their fuel reserves are barely adequate for the advertised delay, but with careful, precise fuel management they will be able to hang on and avoid a costly diversion to the alternate.

You have noticed that each of these crews has a common problem. Each needs to stretch its available fuel supply to the maximum in order to safely complete the assigned mission. Each will learn some lessons in the process and may even pass them on to trade magazines or service publications or safety professionals. In the end, each provocative experience will remain captive to its own limited audience. Naval aviators will read about their F-4 compatriot in some official military publication. Business and executive pilots will follow the G-II flight in a trade magazine. Airline crews will skim the incident reports but there will not be any crossfeed of operational information. In the normal course of events three private lessons on fuel conservation will remain private.

Aviation is an unusual profession. We all share the same sky, the same energy sources, the same airports and the same risks. We learn useful lessons and make preventable mistakes, but we don't talk to each other. We are a fractured family with common needs and little communication. "Crossfeed," my personal column in *Business and Commercial Aviation*, was my attempt to bring that family together in 1,200 words each month.

It was a big family and we had good times together, learning and laughing at the same time. Bush pilots, jet pilots, helicopter drivers, demo pilots, Americans, Canadians, Frenchmen, Italians and more, all contributed to the fellowship with their personal experiences and interest. And each family reunion—each "Crossfeed" column— emphasized one operational challenge. Hydroplaning, weather patterns, air-traffic control, wind shear and medical considerations were all discussed at one time or another. Those meetings always adjourned after 1,200 words, about four minutes of reading, when the subject at hand was retired. And each column was prepared to leave the reader—Ensign, Major or Airline Captain— with some specific operational knowledge that could be used on the next trip. "Crossfeed" was a page for the airborne working stiff.

Eventually, those family meetings accumulated a body of information for every serious aviator. This volume uses those original "Crossfeed" pieces as a basis but there is much more. Twenty-four years of professional flying have given me a unique vantage point for sharing the experiences of many friends, some who survived and a few who didn't. God has given me an ear for stories. Jason Aronson, Inc. has given me these pages. Together we give you this book. Enjoy!

HUMAN LIMITATIONS

We don't like to admit it, of course, but there are real limits to what human bodies can endure. Some limits are always present, such as reduced consciousness at 40,000 feet. Some are limits of time and aging such as presbyopia, the inevitable impairment of vision with age. In all places and at all times, the wise aviator will consider his own personal limits along with those of his airplane and equipment. Consider the case of one who didn't.

Our primary mission in Helicopter Antisubmarine Squadron 5 aboard the USS Lake Champlain, was to track and destroy enemy submarines. In those halcyon days of the early sixties there was much tracking, no destruction. It was a cold-war Navy. Each individual mission lasted four hours and was normally flown with three other HSS-IN helicopters, the Navy's version of a Sikorsky S-58.

During the winter months we were required to wear an evil device of subtle torture known as a "poopy suit," a name that was never fully explained. The poopy suit was made of rubberized canvas and was entered through a slit in the chest. Tight rubber cuffs pulled over the head and wrists and rubber boots were welded to the bottom. It was a hot/sweaty, cold/clammy encumbrance without any appropriate openings for comfort or convenience. Maybe that's why they were called poopy suits.

Now ponder for a moment the traditional Navy emphasis on coffee. Life aboard ship is a little like life in prison with the added risk of drowning, so coffee became a very important pastime. Without families, women, liquor, TV, or radio, the coffee pot became a focal point of life. There was little else to do.

Consider also the peculiarities of our particular mission. Long-duration helicopter flights, often in cold weather with cockpit temperatures well below freezing. (There was no heater.) High levels of vibration from a 1,500 horsepower four-banger at your feet and a primitive rotor system at your back. Significant levels of stress—stark raving fear at times—as we developed the first all-weather, day/night helicopter tactics.

Now mix all those elements together and you can appreciate Lieutenant Jenning's predicament. Too much coffee, vibrated into his cold-shriveled bladder during three hours of terrifying night instrument flying in the North Atlantic, in a poopy suit without a drain. As the fourth hour begins he knows that the battle is lost, and he breaks radio silence:

"Canvas Leader, do you read Canvas 52?"

"Go ahead, Canvas 52."

"Canvas Leader, Canvas 52 has a problem and is returning to Homeplate."

"Roger, Canvas 52, and state the nature of your problem."

"Uh, we've developed some high pressure here and we don't think it will hold another hour."

"Understand 'high pressure,' 52. Is that oil or hydraulic?"

"Uh—it's—we—blztr prl chxt—"

"Say again, 52."

"Fifty-two is . . ." (transmission fades away to garble again).

"Fifty-two, this is Canvas Leader. Please state clearly the nature of your problem."

"Okay, Leader, we are returning to Homeplate because—I—have—to—TAKE—A—LEAK."

"Roger, 52. I hope everything comes out all right."

Sounds of snickering on the discreet UHF channel and then silence.

Clearly Lieutenant Jennings had exceeded a basic human limitation and paid the price. Fortunately, in his case, everything did come out all right, but many human limitations are not so forgiving.

Alcoholism: The Booze News

The entrance to the airline's flight operations office at John F. Kennedy International Airport is in the back, off the building next to the Dempster Dumpsters. In fact, the setting is more appropriate to a Mickey Spillane novel than to the operation of a major air carrier. The stairwell to those second-story offices is twenty-five years old and looks it. There is no glamour reporting to work here, especially at night when the single outside light merely adds to the dingy surroundings.

After seven or eight years, that route from the parking lot, past the Dumpsters and up the stairwell, had assumed a rigid familiarity, so that anything new or different stood out. This particular trip was an evening departure, and when I arrived the ground floor was deserted. Nothing new. The stairwell, however, smelled like a broken bottle of gin, and that was definitely different.

Now you have to understand that liquor on airline property is like ham at a bar mitzvah. You could easily sneak it in if you had a mind to, but it is never allowed. Still, there were enough gin vapors in that stairwell to constitute an explosive hazard. It was a puzzle until we were halfway to Chicago.

Over Dunkirk VOR, when the captain leaned toward me to reach the transponder "ident" button on the overhead, the puzzle was solved. This old geezer had braced himself with a snootful for that long night trip.

Alcohol is a significant problem at all levels of aviation. The parking lot at work, which is used by pilots, flight attendants, mechanics, managers, foremen, cleaners and all, often contains an impressive assortment of empty bottles. Alcohol is also so much a part of our culture that criticism is normally benign or apologetic. After twenty-four years in the cockpit I have developed some

3

very strong prejudices about drinking and aviation, and I think those prejudices are justified.

General aviation fatalities in the United States over the past years indicate an alarming incidence of alcohol involvement, and these statistics only refer to the positive presence of alcohol in the pilot's body as revealed by an autopsy. Unfortunately, no one knows the real extent of alcohol's effect on aviation safety because of the insidious nature of the problem.

Alcohol is a drug that acts as a depressant on the central nervous system. As such, it markedly alters the activating and inhibiting functions of the brain. The results are well-known—lowered blood pressure, increased reaction time, slurred speech, loss of coordination and modified behavior.

Alcohol is metabolized by the liver at a fixed and constant rate, equal to about 0.015 percent each hour. It doesn't matter how much alcohol is in your bloodstream, how much you exercise or how much coffee you consume afterward. The metabolic rate is a steady, plodding 0.015 percent per hour, and you cannot change it with aspirin, oxygen, coffee or a "hair of the dog."

Suppose you stop in for a toddy on the way home and drop two Martinis in the first hour. Your blood alcohol will approach 0.10 percent, the level set by many states as the limit for drunk driving. Two more Martinis in the next hour and you head home for nine hour's sleep in preparation for tomorrow's trip. Your blood alcohol will likely be in the 0.20 percent category.

Nine hours of sleep, shower, shave and coffee and you feel like a million bucks. Well, at least you look like a million bucks, and your blood alcohol is well below that legal limit of 0.10 percent. Remember, however, that that definition of impairment is pertinent to motor-vehicle operations and bears little relation to flying.

One study of the effects of alcohol on experienced pilots utilized single-engine aircraft under simulated instrument landing approach conditions. Some of the subjects had distinct difficulty at a blood alcohol level of 0.02 percent. There just is no acceptable level of blood alcohol compatible with flight.

If you went to bed at 2300 hours with an alcohol level of 0.20 percent, it will still be 0.06 percent as you brush the fuzz off your teeth at 0800. When you rotate that great silver bird at 1000 hours, your blood alcohol will be sufficiently high to cause some measurable impairment of the very skills on which your passengers' lives now depend. You have satisfied every legal requirement, and yet those toddies from the previous evening are very effectively reducing your performance. Even when all the alcohol is metabolized, the hangover effect will getcha for several more hours. Hangovers not only feel bad, they are a very real medical phenomenon. In fact, several physiological mechanisms contribute to the blah sensation that follows excessive drinking like a shadow.

One major effect of alcohol is a temporary alteration of the fluid balance. In short, the sauce dehydrates your body. It accomplishes this by stimulating the kidneys to produce a large quantity of diluted urine so that the body loses more liquid than it takes in. This dehydration produces a concentration of all those solutes normally found in bodily fluids, and that chemical concentration causes weakness, fatigue and irritability.

Another major element in hangover production is the assortment of organic impurities found in all alcoholic beverages. These aldehydes, ketones and so on are metabolized in complex ways and may remain in the bloodstream long after the alcohol is gone. While present, they produce fatigue and the blahs of their own making.

You can minimize your intake of these impurities by drinking only clear beverages such as gin and vodka, but you cannot eliminate them altogether. There is a real germ of truth in that old invitation to "name your poison."

Just when you need extra rest to compensate for these deleterious effects on the old bod, the alcohol itself will have degraded the quality of your sleep. Specifically, alcohol in the blood reduces the proportion of sleep spent in the dream state. This reduced dream sleep has been shown to induce fatigue, anxiety and impaired concentration.

Finally, the minute residual alcohol level in your blood will markedly affect the vestibular organs that control balance and spatial orientation. Your adaptive resistance to motion sickness and disorientation may be compromised sufficiently to cause mild nausea and discomfort. Vision may be impaired, especially under low illumination levels found at night or in clouds. Both of these problems can be induced or aggravated by rotational and G forces for periods as long as 20 to 36 hours after drinking.

What about a cure for the hangover? The very best one is to allow a maximum amount of time for sleep and recuperation between drinking and flying. Aside from time, there is no magic medicine, but you can minimize the more obvious effects and maximize your performance with a few simple remedies:

• Stick to vodka or gin to reduce your intake of impurities.

• Reduce dehydration by drinking several glasses of water before going to bed and several more when you awake.

• Try some coffee or tea. The caffeine is a stimulant and will tend to counteract the depressant effects of alcohol. Your liver will not metabolize any faster, but at the very least you will be a wide-awake drunk.

• Eat some fresh fruit or honey, or drink some fruit juice. The fructose sugar in fruit and honey is believed to help the body to eliminate alcohol. Besides that, you need to eat something even if you don't feel hungry.

Bear in mind that none of these remedies will cure the hangover. At best you can hope to mask the more unpleasant effects of overdrinking.

Don't kid yourself about those Martinis. They will have a specific and measurable effect on your performance for much longer than most of us would care to admit.

The important message is that alcohol dependency is as large a problem in aviation as it is in the general public despite any wishful thinking to the contrary. Also, that excellent help is available to detoxify the alcoholic and terminate his physical dependence if and when he is ready to seek help. Psychological and emotional dependence may be more difficult to arrest, but it can be done. In a word, when you face that difficult personal task of coping with an alcohol-dependent friend or associate, treat him like a man. But never back down on the problem. If you suspect that you may have a problem yourself, get help—fast. Alcohol dependence can be whipped, but only with professional help.

Maryjane Is No Lady

2

Since those more innocent days of the early sixties, other recreational drugs have become popular. Even after the demise of the heavy drug culture, there is an underlying fascination with chemohappiness, which creates the climate for social tolerance of drugs. One of those drugs is particularly available, socially popular and has, unfortunately, been officially declared benign by a badly mistaken national commission. The fact is that marijuana is a demonstrably harmful drug with particularly serious consequences in the aviation environment.

Call it whatever you like. *Cannabis sativa*, pot, grass, maryjane, dope or weed. Marijuana has become an established element in our youth culture with considerable spillover into the adult population. That spillover is the result of the ready availability of the drug and the natural process of time, which has seen the youths of the sixties grow into the adults of the seventies.

Ironically, general use of marijuana has been subtly fostered by the National Commission on Marijuana and Drug Abuse. This government agency released its opinion that pot does not appear to kill or cause brain damage, does not cause birth defects, is not physically addictive and does not lead to crime or the use of stronger drugs. The commission recommended the repeal of jail terms and fines for private pot smoking. Many have heralded those conclusions as the definitive statement on marijuana safety, but there are conflicting opinions.

The World Health Organization consistently has warned against the use of marijuana, and the British and Canadian governments have issued clear warnings about its use. Several recent studies have substantiated earlier claims that marijuana is a potent and potentially harmful drug. As a minimum, its use should not be tolerated in aviation. That much is absolutely clear.

Marijuana is an unusual drug. The active ingredient, tetra-hydrocannabinol (THC), is retained in the body for long periods. One study indicates that 30 percent of the THC is retained in the body after seven days. Moreover, it is eliminated at a much slower rate than is the first 70 percent. The result is that with regular use THC accumulates in the body.

Its effect on the individual is even more interesting than those stark percentage rates, and its use defies any direct comparison with alcohol.

Alcohol is water soluble. If you have a few drinks today, all the alcohol will be purged from your body within a maximum of 36 hours since it is metabolized like any other food. It leaves the body promptly and without residue.

But THC is a fat-soluble compound and that makes all the difference. The THC remains in the fatty material of body cells for long periods and, with regular use, accumulates there. The chemical is changed only slightly by metabolism, and even then it may be converted into a more psychoactive form. There are about 50 different cannabinoids in marijuana, and they appear to have an adverse effect on all body cells. Researchers have documented medical changes to the membrane of brain cells, to the red and white cells in the blood, and to liver and lung cells. In aviation, there is a special reason to be concerned about the effects of THC on brain cells.

Brain cells have fine, hair-like extensions that are used to communicate with other brain cells. These thousands of synaptic structures are, in fact, the network for brain activity, and as such they are the mechanisms of the mind. Marijuana damages those vital connectors.

One study at Bristol University involved ten young marijuana users who manifested noticeable behavioral changes. X-ray examinations of their brains revealed that they all were suffering from cerebral atrophy, a sort of slow wasting of the brain tissue. The degree of atrophy was in proportion to the duration of marijuana use.

Another study exposed monkeys for six months to doses of marijuana that corresponded to moderate to heavy human use. Then they were taken off the marijuana for eight months before being medically examined. The examination revealed that the location and degree of atrophy in the monkeys' brains were nearly identical to that in the ten young men. Also, the synaptic structures had been altered.

Marijuana's marked effect on the brain's structure is significant, but even more significant is the subtle and frightful effect of the weed on behavior.

Marijuana has a definite adverse influence on the performance of precision jobs. Regular users suffer from distorted emotional responses, disorganized thinking, apathy and indolence. Many of these individuals may suffer from chronically altered judgment, although they appear normal.

The regular marijuana user is prone to error, has difficulty remembering details and often cannot think practically about the future. These effects show up gradually, and they appear to be reversible in the short-time user. As exposure to the drug continues, however, recovery is less and less complete, even after long periods of abstinence.

Studies of the influence of marijuana on driving ability have indicated a reduced ability to judge distance, speed and road conditions. Because the drug leaves an accumulative residue, driving performance is impaired even between uses. The regular marijuana user is never completely free of marijuana's influences.

The mental transformations of the marijuana user are gradual and subtle. There may not be any obvious sign of impaired ability, but it is there nevertheless. Fortunately, there are medical tests to determine marijuana use.

Urine tests can indicate pot use within the preceding 24 hours. More difficult and expensive tests of blood, fat or feces can be used to measure the average level of intake over a period of several months. Any defensiveness about the use of marijuana or other drugs raises a red flag. The very act of sympathy toward the use of drugs by anyone involved in aviation suggests a serious lack of judgment at the very least.

Marijuana is a mood-altering, mind-changing drug that seriously affects an individual's suitability to operate or maintain sophisticated equipment. In aviation there is as much reason to restrict the activities of the known marijuana user as to limit those of the alcoholic. There is just no place for the pot user in our industry.

Sinking Spells

3

Alcohol and pot can compromise your flying with fatigue, disorientation, forgetfulness and the blahs. So can lack of sleep, bad food and overwork. Physical rest, in fact, inevitably becomes a major priority in serious flying.

The junior officer bunkroom reserved for my squadron's pilots aboard the USS *Lake Champlain* had eighteen bunks, two sinks and one shower. There were no "weather openings"—windows, to the landlubber—so that all illumination was artifical. In actuality there was never much light in our room because someone was always sleeping after a mission or napping before one.

That dark environment created a problem because when you awoke, it was never clear whether the time was A.M. or P.M. Now, Fred McIntosh solved that little ambiguity with ingenious simplicity. When he slept during the daytime he left his socks on. When he slept at night, he slept barefoot. Fred was no dummy about organizing his clock, and those demands for catnaps do not change in a flying career.

The pilot lounge in Chicago includes a so-called "darkroom" for sleeping. (Incidentally, the sign on that door used to read, "Pilots Only" until the advent of cockpit ladies, so management has predictably hung a new sign that reads "*Male* Pilots Only.") At any time of day or night you can see the world's finest pilots stagger out of that dark room back to the bright lights and blaring TV. More often than not you will find that those nappers are co-pilots because they know their inflight rest will be minimal.

Captains are different, especially long-range captains, and most especially at night. Coast-to-coast freighter trips normally depart at some miserable hour of gloom, so that the first topic of conversation after level-off is who naps when. If you are a co-pilot, you naturally defer to the captain; besides, he's older and needs more sleep. If you are a captain you graciously accept that offer of first sleep and the pattern is set: the captain reclines his chair to study the instructions inside his eyelids. The flight engineer pretends to be alert, but you can never trust an FE to stay awake even in broad daylight. The co-pilot tends the store. If that co-pilot is timid, and if the captain has spent the day landscaping his yard, there will only be one napper. That is why so many users of the "darkroom" are co-pilots.

Sometimes in the cockpit there is more than one napper at a time. Sinking spells are a fact of life but their results can be particularly serious in aviation. Consider:

A 707 freighter, passing Las Vegas at the end of an all-night KFG-LAX nonstop, continues westbound at FL390 (30,000 feet above sea level), ignoring the repeated descent clearances issued by LAX center. After considerable effort and anxiety, radio contact is established through Arinc when the aircraft is nearly 100 miles west of LAX. Fortunately, the SELCAL (Arinc selective callup) chimes are loud enough to awaken at least one of the three sleeping crewmembers, and there is enough fuel for a safe return to LAX. In another instance, a DC-6 crew spent half an hour circling Atlanta when they all fell asleep with the autopilot on and the turn knob mistakenly set for a very gradual turn.

Cockpit slumber parties are one of the more dramatic effects of severe fatigue. Some will say that such conduct is inexcusable. I wonder if that oversimplification serves to mask some very real physiological problems. Several things can contribute to those awful sinking spells, but some of these factors can be controlled.

One hundred years ago, crosscountry travel was by horse or train at speeds that never exceeded 25 miles per hour. Across the United States in those days there were at least 300 different time zones because each city and town set its own clocks at local high noon. Around the world, the concept of 36 designated time zones was as foreign as the space shuttle and electric toothbrush. The science of transportation simply had not progressed to a point that necessitated standardized time keeping.

Today, abnormal working hours and rapid time-zone changes are practically inherent in an aviation career. There may be no way to avoid them, but it is best to understand their substantial effects on the body.

Physiologists have identified more than 100 body systems that conform to a 24-hour biological clock. They refer to this daily cycle as the "circadian rhythm" because it takes place once a day. Many of the body's most significant systems have a measurable 24-hour period:

• Temperature reaches a maximum between 1400 and 1700 hours and a minimum at about 0500 hours.

• Several types of body fluid are secreted at nearly fixed times to create a tidal ebb and flow that are easily observed and measured.

• Intellectual efficiency—the ability to think clearly—has been shown to be lowest at approximately 0400.

• Blood circulation, respiration, liver activity, blood composition, some cellular processes and even mood are all linked to a free-running, 24-hour biological clock that is resistant to change.

• Sleep has its own measurable stages that are broadly categorized REM (rapid eye movement) and nonREM. The two phases, REM and nonREM, last about 100 minutes. An eight-hour sleep would contain roughly four to five such cycles, although their makeup changes markedly through a typical eight-hour sleep. We spent 90 percent of the first four hours in nonREM sleep, while in the last four hours there is a much greater proportion of REM sleep.

Any short sleep cuts off a disproportionate amount of REM sleep, with measurable side effects: irritability; short-term memory loss; and impaired judgment and decision-making capability. Short naps cannot restore REM sleep, which we must have to function properly. Eight-hour sleep periods are the only way to maintain the REM sleep supply.

Any flying schedule that interferes with normal wake-sleep cycles will have a noticeable impact on your performance. In effect, your innate 24-hour biological clock will be desynchronized. Symptoms will vary with age, sex, physical conditioning, circumstances of the trip and your past travel experience, but you should expect any or all of the following:

• Fatigue, insomnia, headache, blurred vision, dizziness, emotional depression and mental confusion in proportion to the disruption of your normal sleep.

• Minor discomfort on any trip that crosses two or more time zones in a single day or which alters your wake-sleep cycle by that much.

• More intense discomfort for trips through four or five time zones in a day; ideally, a full day should be devoted to rest after arrival.

HUMAN LIMITATIONS

You may be young, tough and resilient, but you are not immune to circadian desynchronosis; we all get it and cope in individual ways. Here are some measures that can minimize the effects:

• Intelligent planning of the itinerary can go a long way toward preventing the difficulties in the first place. In most cases you will not have the luxury of selecting the departure time, but when you are involved in trip planning, recommend travel times that will minimize physiological impact. Daylight or early-evening travel is invariably best. Stops and layovers ought to be long enough to provide effective rest periods.

• Adequate rest before the trip is a prerequisite to minimizing circadian disruptions. Schedule your own time accordingly, and schedule other pilots with a critical regard to their individual rest needs.

• Prepare early. Really arduous trips, especially those that will begin at unusual hours, should be carefully flight-planned and prepared in advance. Some last-minute changes may be necessary because of weather and route availability, but major choices of fuel loading, alternative airports and route can be tentatively settled the previous day.

In flight there are several measures that can help you through those long backside hours:

• Raise your feet as much as possible to prevent blood from pooling in the legs. Stand up at least every two hours to revitalize circulation.

• Eat lightly. Heavy food (or too much caffeine) will aggravate the problem.

• Breathe 100-percent oxygen for a few minutes every two hours.

• Maintain the lowest feasible cabin altitude. Recent research indicates the circadian desynchronosis intensifies with cabin altitude.

• Consider using bright lights in the cockpit, even at night (provided you are at IFR cruising altitudes where there are no traffic conflicts), because there is evidence that increased white light increases alertness.

• Recommend that your passengers travel with a companion. No one is sure why, but solitary travelers take longer to adjust to time-zone changes than those traveling with companions. When appropriate, tell the boss that it could be beneficial for him to travel with an associate.

• Flight crews should be aware of each other's limitations and consider brief individual naps during periods of low activity to ward off fatigue during approach and landing.

SINKING SPELLS

It doesn't matter whether you call it lousy working hours, circadian desynchronosis or that over-simplification "jet lag." When you fly at other than normal working hours, you should expect to function somewhat below your best.

Lack of sleep is the most obvious cause of pilot fatigue but it is not the only one.

HYPOGLYCEMIA

Another factor to consider is *hypoglycemia*, or low blood sugar, which can cause malaise, fatigue, disorientation and even lapse of consciousness. It is controllable with some very simple dietary tricks, but it helps to understand the mechanisms involved.

When you awake in the morning, your blood glucose level will be low from the overnight fast. If you start with coffee and sweet rolls, or any other highly refined carbohydrates, you may induce reactive hypoglycemia.

It works like this: your system converts refined sugars and starches into glucose so rapidly that your blood sugar level rises at an abnormal rate. When the homeostatic system that balances glucose levels senses the sharp rate and rise, it signals the pancreas to release insulin proportionately. In this case, the rate is abrupt and can only be sustained over the very brief time it takes for your body to convert the refined carbohydrates to glucose. The end result is that too much insulin is triggered to the bloodstream, so that your glucose volume is soon driven well below the original, fasting level.

Proteins from a normal ham-and-eggs breakfast are processed by the body at a much more steady rate so that appropriate quantities of insulin are metered out to stabilize glucose levels at the optimum point. Protein reduction and conversion continues for several hours and eliminates the peaks and rebounds induced by pure carbohydrate intake. Four dietary practices will prevent reactive hypoglycemia:

• Avoid refined carbohydrates (sugar and all refined starches).
• Eat protein-rich meals every four hours, especially when on flight duty.
• Use fruit or protein snacks for pick-me-ups at odd duty times.
• Substitute milk or fruit juice for coffee and soft drinks.

And speaking of coffee, I used to fly with a guy who drank gallons of coffee and never could stay awake. Turned out that he suffered from *caffeine toxicity*, a not-uncommon problem that can cause poor sleeping, nervousness, headaches and lethargy.

HUMAN LIMITATIONS

Coffee, tea and cola, in moderate amounts, promote quick energy and clear thinking through the stimulant effect of caffeine. Above a certain level, caffeine ceases to be beneficial and becomes a hindrance to normal functioning. One cup of coffee or tea contains about 100 mg of caffeine. Twelve ounces of cola contain about 50 mg. Some doctors feel that 500 mg per day is enough, 750 mg is questionable and 1000 mg addictive, and possibly dangerous.

Check your caffeine intake and be sure to include all possible sources. Coffee, tea and cola are obvious contributors, but caffeine is also present in chocolate and in many of the nonprescription headache and cold medicines, and over-the-counter stimulants.

Pilots of pressurized airplanes should consider the problems induced by the *very dry cabin environment*. This extremely low humidity can affect your physical well-being because of imperceptible water loss through perspiration and respiration. If this water is not replaced, the respiratory tract and eyeballs suffer, and a feeling of weariness may develop. Insensible water loss at altitude, in a pressurized aircraft, is about 40 percent greater than normal. The solution is to increase your water intake at least two ounces above normal for each hour of flight. Coffee, tea and cola don't count.

Now about those *cigarettes*. One smoke raises the carbon monoxide in the blood to a level that equates to a state of hypoxia at 7,000 feet. Two cigarettes smoked consecutively raises the level to 10,000 feet, and these levels are further aggravated by actual cabin altitude. Smoking is unquestionably a contributor to fatigue.

When you do get drowsy inflight, try a few exercises, eat some nuts, suck a little oxygen and try to recall the thrilling contents of this book. If that doesn't keep you awake, nothing will.

Decompression Sickness

4

Speaking of physical limitations, consider the possibility of decompression sickness—the bends—at medium to high cabin altitudes. I have a graphic memory of the underlying cause of the bends, expanding gases.

The decompression chamber at Pensacola Naval Air Station looked like a primitive submarine with thick steel walls and heavy, bolted portholes. Ten of us aviation cadets and a single Navy nurse trainee filed into the chamber along with an instructor. As the instructor briefed us about the chamber ride he pointed out a limp surgical glove hanging from the ceiling, a glove which we would observe to expand from the effects of reduced ambient pressure as the chamber's altitude was steadily increased to 30,000 feet.

We are ready now, and there is a great hissing and sighing as air is pumped from the chamber. Lots of idle chatter over the noise as we approach 10,000 feet. At this altitude, the rubber glove is just nicely shaped from the expanded air inside.

Soon after the oxygen masks are put on I sense that certain interior organs in my lower abdomen are duplicating the glove's expansion, and it becomes a contest to see which will burst first.

At 20,000 feet the glove looks like a grotesque, swollen salami with five fat sausages attached. The gases inside have swollen incredibly in this reduced ambient pressure, as have my personal gases, which are now screaming for release through the exhaust valve on which I am sitting.

Now consider the scene. Ten crass aviation cadets, normally willing and ever eager to relieve the pressures of flatulence in one another's company as an element of pure macho humor, but now restrained in deference to that single female presence.

17

And that poor girl. I mean female gases respond to Boyle's Law in exactly the same fashion as male gases, even if female sensibilities are somewhat more civilized, so that eleven of us sat there for one last moment of ultimate restraint before the inevitable explosion.

When it did come, as it must, that eleven-part chorus was largely obscured by the hissing, sucking vacuum pumps of the chamber, but the rumbling vibrations were readily transmitted along those crude wooden benches. I could see nine other male faces contorted in laughter behind their oxygen masks. My classmates—and I— were enjoying the crude humor of the moment. Our female companion was not. Her face was a great blotch of red blush between auburn hair and olive drab mask. Expanding gases had had different but predictable emotional effects on all of us in that little chamber.

There are other, more serious effects of expanding gases that may be new or unfamiliar to many serious pilots, particularly with the recent advent of higher flying planes.

Turbocharging is enjoying a rebirth in general aviation after years of neglect by the military and airlines. The benefits are obvious: improved climb performance, vastly increased ceilings, greater speed and mechanical simplicity. The risk of hypoxia (a deficiency of oxygen) is nearly eliminated by installing or carrying onboard a simple oxygen system to compensate for the reduced pulmonary partial pressure caused by rarefied air at higher altitudes. But pilots who anticipate unpressurized, turbocharged operations above 18,000 feet should understand the hazards of decompression sickness, or "the bends."

A 45-year-old military pilot departed Randolph AFB on a morning flight in a T-33. The aircraft climbed rapidly to FL 350 with the cockpit pressurized to an altitude in the mid-20s. One hour and twenty-five minutes after takeoff the pilot complained of tingling in the arms, light-headedness and uncoordination of the left hand. He turned his oxygen regulator to 100 percent, relinquished control of the airplane to the pilot in the rear seat and requested that the heat be turned up.

After landing, the pilot was ashen, perspiring and breathing heavily. He had to be helped from the cockpit. Rushed to the base hospital, he was admitted with a temperature of 101°, slurred speech, left facial paralysis and parasthesia of the left arm.

During the next three weeks of hospitalization the pilot made some progress and was released for convalescence at home. Two months later, however, he continued to complain of numbness, weakness, confusion and uncoordination. Four months after the incident he was medically retired, and 18 months later he was still afflicted with these problems. Decompression sickness, though it had occurred at medium altitudes, had done permanent neurological damage.

A 36-year-old military pilot was summoned for a night mission after finishing a supper that included two cocktails. Eighteen minutes after takeoff, at a cabin altitude of 22,000 feet, he told his radarman he felt sick. Seven minutes later he reported severe chest pains and subsequently slumped over to one side of the cockpit, unconscious.

After a successful landing by the radarman, the pilot was found to be cyanotic, or bluish-colored, cold, clammy and irritable. When admitted to the hospital a few minutes later, he was perspiring, confused, flaccid and uncoordinated. Despite constant, expert medical attention, he died 12 hours after the first signs of trouble. Decompression sickness had done fatal damage to the nervous system.

A military transport, cruising at 25,000 feet, was deliberately depressurized to check a door lock. After 45 minutes of unpressurized flight, two crewmembers developed pain in their knee joints. Both men had been using 100 percent oxygen throughout the unpressurized portion of the flight, and pain disappeared when the airplane was repressurized.

The body is saturated with nitrogen and other inert gases, and each tissue and organ contains a partial pressure of gas equal to that in the surrounding atmosphere. When ambient pressure is reduced, as it is in unpressurized flight, there is more nitrogen pressure in the body than in the surrounding air. At nominal cabin altitudes—below 20,000 feet or so—that pressure difference is inconsequential for a young, healthy adult. At higher cabin altitudes, especially on longer flights, nitrogen will begin to bubble out of the body's tissues, pressing on nerves or joints and causing discomfort, pain, neurological problems—even death. In the lungs, those nitrogen bubbles impair breathing in a decompression symptom called "the chokes."

Pilots and operators who anticipate unpressurized flights above 18,000 feet should know that:

- Decompression sickness is not caused by oxygen deficiency

and cannot be avoided by oxygen use, with one exception. There is evidence that continuous use of oxygen will gradually wash some nitrogen out of the body after several hours.

• Decompression sickness is caused by a decrease in ambient pressure, and the one positive remedy is to increase that pressure as quickly as possible. In most cases, an immediate descent and landing is required.

• Scuba diving within 24 hours of flying will significantly increase the potential for decompression sickness.

• Physical exertion immediately before or during flight will increase the possibility of physical decompression.

• Any alcohol consumption within 24 hours of flight will facilitate the body's tendency to release nitrogen bubbles under reduced pressure, thereby increasing the risk of decompression.

• Decompression sickness should always be treated as a serious medical emergency requiring an immediate landing and professional medical care.

Pilot Inflight Incapacitation

5

Decompression sickness can be serious when it occurs although it is, in fact, quite rare in aviation, particularly general aviation. Still, as aviation hardware and procedures improve, obscure safety problems become more apparent. For example, the whole human element is beginning to receive the attention it has always needed.

Physical collapse or impairment in the cockpit occurs more frequently than many of the classic mechanical irregularities.

Consider the following excerpt from a routine airline safety bulletin:

During a single 12-day period we had several cases of incapacitated pilots.

In the first incident the captain suffered an attack of violent nausea and stomach cramps. This occurred in the vicinity of Hill City while westbound to LAX in a B-727. The captain elected to occupy the forward jump seat. He directed the first officer to continue to LAX, as opposed to landing short, and to make a coupled approach. The second officer handled ATC and company communications; he occupied the captain's seat for the approach and landing.

Another incident involved subtle incapacitation of the captain.

The first officer, who was flying, and the second officer noticed that the captain was having difficulty verbalizing. He could not complete his radio transmissions. This occurred when the captain initiated several unnecessary calls and again when responding to ATC.

The captain was unable to respond to a suggestion that he take oxygen and/or a stroll in the cabin. The second officer had the first flight attendant inquire if there was a doctor onboard. There wasn't. The captain's physical appearance was normal; the major

21

symptom was inability to verbalize and trouble with his thought process. The first officer decided to land at ORD.

Aviation medical specialists divide such incapacitations into two broad categories—obvious and subtle.

Obvious inflight incapacitation is that sudden functional loss you would expect from massive heart failure or gross cerebral stroke. An example would be the collapse of your flying partner, who slumps unconscious with little or no warning, leaving you to fly the airplane, remove any interference caused by his inert form and execute a safe landing. A far more insidious predicament is created by the pilot who passes inconspicuously into a condition of dysfunction.

Subtle (or partial) incapacitation is that situation in which a pilot appears to be functioning but is mentally disengaged from his immediate responsibilities and surroundings. He may sit with eyes open and hands on the controls, looking perfectly normal to his fellow crewmembers, while his cerebral cortex, that portion of the brain responsible for thought and analysis, has ceased to function.

There are several potential causes of subtle incapacitation, but two are of particular interest to the pilot.

Hypoglycemia occurs when the regulating mechanism that controls the blood sugar level malfunctions. The brain is deprived of its only nutrient, glucose, and sweating, weakness, tremors, mental confusion, irritability, headache and even convulsions can result. In fact, by starving the brain, hypoglycemia can mimic a neurologic or psychiatric disorder.

Only your doctor can properly diagnose your individual susceptibility to hypoglycemia, but there is a simple preventative: proper diet. Highly refined carbohydrates such as sugar and white flour can trigger a low blood sugar response, especially when taken on an empty stomach. The coffee and doughnut breakfast is as potentially dangerous as bald tires or a low battery. Particularly when flying, concentrate on high-protein snacks such as nuts and cheese, eat fruit in place of sweets and consume a protein-rich meal every four hours, if possible.

Cerebral dysfunction, another cause of subtle incapacitation, is an inclusive term used to describe any alteration in normal cerebral function. The cerebral cortex, which is responsible for what we call human intelligence, generates those intricate mental processes that allow such complex psychomotor activities as flying. Mild cortical defects are not uncommon and may go undetected indefinitely if the central nervous system is not stressed beyond routine levels. If a pilot who is suffering cortical defects encounters critical stress during some pivotal flight maneuver, however, he may cease to function effectively at the worst possible moment.

Hypoglycemia and cerebral dysfunction are distinct but not exclusive causes of subtle inflight incapacitation; brain tumor, epilepsy, and even psychological stress can also trigger serious mental lapses.

One of the most thorough operational studies of pilot incapacitation was performed by United Airlines at its Denver, Colorado flight training center. Qualified DC-8 and 737 line crews were utilized while flying the respective simulators. That important study is the basis for a four-step incapacitation procedure now used by nearly every major airline.

The first step is so elementary that it can be overlooked. Assume control of the airplane and (if necessary) climb to a safe altitude. I know that sounds obvious, but your startle reaction may prompt you to exert undue attention to good ole Fred as he clutches his chest and grimaces with intense pain. Avoid that temptation and just fly the airplane. Carefully, even methodically, check the position of essential controls and switches because your partner's behavior just prior to his failure may have been irrational. A thorough cockpit inventory will further serve to nullify your own emotional response.

When you are satisfied that the airplane is safely configured (fuel, hydraulics, generators, anti-ice, autopilot, radios, etc.) ask ATC for a holding vector and consider declaring an emergency. Right then you'll need all the help you can get and the NTSB is going to require a written report anyway because of the incapacitation.

Next, when flight conditions are stabilized you will need to restrain and remove the sick pilot for his own good as well as to reduce the distraction and possible interference he will create. It takes two people to handle the dead weight of an unconscious body without risk of interference to controls and switches, so you will need help. Delegate the first aid to someone else so you can concentrate on your new responsibilities as sole pilot.

When you are alone, reorganize the cockpit for a landing as soon as practicable. You may want someone in the empty seat to read checklists and reach out-of-the-way controls, but he must be carefully briefed on the narrow limits of his activities.

Your primary burden is still to fly the airplane, but when time allows, call ahead for medical help. The really essential element is an ambulance. Doctors are helpful, but your patient will very likely require treatment which is only available in a hospital. If you have time, give a brief factual summary of the patient's symptoms and condition, but don't overload yourself or delay the landing. After the landing is completed, you can lavish your full attention on the patient's care.

The incapacitation procedure can be reduced to four simple steps:

1. Assume control and fly the airplane to a safe situation.
2. Restrain and/or remove the incapacitated pilot.
3. Reorganize the cockpit and prepare for a landing.
4. Arrange for an ambulance to meet the airplane as time and workload permit.

Make it a point to discuss the potential consequences of both obvious and subtle inflight incapacitation with other regular crew members. If possible, plan annual drills on the ground or during simulator sessions. Incapacitation is a statistical reality, frequently included in internal airline operating bulletins. It can happen anywhere.

There Comes a Time 6

Serious inflight incapacitation is a real, although rare, threat to flight safety. Junior airline pilots usually have a cavalier attitude toward the possibility of their captain's inflight demise. Sure it's a tragic, perhaps dangerous, occurrence, but every step up that seniority list is valuable. We used to say as co-pilots that if the captain died during a trip we'd call his widow—as soon as we had informed the keeper of the seniority list. Still, given enough time, seniority is as inevitable as taxes—and impaired eyesight.

I used to fly with a wonderful old guy who loved a joke, and he had a few special ones of his own. When making an approach with a new co-pilot he would announce calmly, as the airplane crossed the runway threshold, "I am closing my left eye." Then, just as his young apprentice began to wonder, he would announce, during the roundout, "I am now closing my right eye." The fact is that none of us co-pilots ever had enough courage to look over there and check at that critical moment, and so no one knew for sure.

One sure thing about an aviation career is that both eyes gradually will become less useful after about the age of 40. Doctors call the phenomenon *presbyopia*, or "old man's vision," and if you haven't felt the effect yet, you will.

When your eyes are relaxed, they are adapted for distant vision. To focus on an object closer than 20 feet, the ciliary muscles contract to increase the curvature of the lens so that it *accommodates* near vision.

Beyond the age of 40, the lens loses some of its elasticity and the eye loses its power to accommodate. When you start holding the approach plate at arm's length to keep it in focus, you can be sure you are a victim of this progressive deterioration. Like it or not, you are a victim of presbyopia and it's time to think about glasses.

HUMAN LIMITATIONS

Eyeglasses are both a curse and a blessing. They help you to see better, but they introduce a whole new menu of problems. My first adventure with eyeglasses was frustrating and unnecessarily expensive, but I learned some lessons.

There are three basic eyeglass solutions to your presbyopia, and each one is a compromise:

Half-glasses, those little granny-style reading glasses that sit low on your nose, will allow you to view close objects through the half lenses and see distant objects by looking over the lenses. Half-glasses are satisfactory if you do not require any distance correction, and many pilots wear them. I never liked my $50 pair because they created a distracting line that constantly bothered me. Scratch half a hundred.

Bifocals are the next logical choice, even if you don't require any distance correction. Bifocals are really two lenses molded into one piece of glass or plastic. There is a distinct line between the two halves, but that line is less objectionable than the top of half-glasses. Properly fit bifocals cost between $100 and $200. My first pair were just satisfactory after I modified them to sit up high enough on my nose. More about that later.

Progressive lenses were introduced within the past ten years. Manufacturers claim that these glasses provide both near and distant corrections with a continuous optical grading or changing of the lens that eliminates the distracting line. My own pair are gathering dust on the dresser because they work properly only when I look through the center 10 degrees. My optometrist and my optician agree that progressive lenses in their present state of development are a distorted mess and a waste of money. I cannot recommend them. If you want to try, the price tag begins at close to $200.

Bifocals are probably your best choice even if you do not need distance correction. You will need to have them made with the dividing line precisely placed at the glare shield level to minimize distraction. That's where you get involved in the fitting process.

First, select a pair of rims and have them fitted to your face. Then have the rims fitted with throwaway plastic inserts cut from cheap sheet stock. Next, sit in the cockpit with your seat and head in the normal flying position and mark the phony lenses with a felt-tip pen where the glare shield bisects your vision. When you return to the optician, he will use that mark to position the bifocal line. It's the only way to get that critical measurement right the first time.

THERE COMES A TIME

When you finally order the glasses, you will find more options than on a new car. Your first decision will concern glass or plastic lenses. Plastic is lightweight and unbreakable but far more susceptible to scratches. Glass is heavy, nearly as durable as plastic, and almost impervious to scratching. I like the plastic lenses because they are so incredibly light. And with a minimum of careful handling, mine are still free of any serious scratches after three years.

Your next choice will be whether to select those photochromic or photoreactive lenses that darken and lighten to compensate for ambient light levels. Don't.

Photochromic lenses are impregnated with micro-crystals of silver halide. Under normal conditions, ultraviolet light from the sun darkens the halide to a neutral gray sunglass in as little as 60 seconds. Unfortunately, cockpit windows will screen out most of those ultraviolet rays so that your glasses will never darken properly. Worse, however, is the fact that photochromic lenses require more than five minutes to lighten. If you fly from bright sunlight to heavy cloud, your photochromics may be too dark for several minutes. Also, they are never completely clear, so that night vision will be hampered to some extent.

One parting word on sunglasses: buy quality. Cheap sunglasses can do all sorts of nasty things, from fatiguing your eyes to distorting your vision. Good sunglasses will filter out the correct amount and type of light, protect your eyes and protect your night vision for the portion of your trip in darkness. If you are not sure about the pair you've been wearing, have them checked by a professional.

Visual Illusions 7

Visual impairment with age is one predictable form of human limitation. Visual illusions, on the other hand, are another form of limitation without age restriction.

When I was a 727 co-pilot at the age of 28, I could read the eye charts down to 20/13 with both eyes. My headlights were really sharp then, and they normally served me very well. My come-uppance was in a rain-shower, on final, at Birmingham.

The approach to Runway 5 at Birmingham is through a notch in the hills that distorts the view of that runway in such a way that hard landings are not uncommon. With rain on the windshield and a minimum of jet experience I was a setup for even the slightest visual illusion.

I wish now that I had a picture of that touchdown at the moment of impact. Technically, it was not a touchdown at all. I'm not sure if there is an appropriate aerospace term, but if that photo appeared in *MAD* magazine it would be captioned "THWACK" or "CRUMP" or "KABOOM" or all three. I am sure that every occupant of that airplane had valid claims for a damage suit because you could hear the vertebrae compressing along with the oleo struts. They all were innocent victims of my visual limitations.

Remember Flip Wilson's girlfriend Geraldine and that one-liner she always delivered after sashaying into view, hands on hips, with a come-hither look? "What you sees is what you get" was her stock opener, although in Geraldine's case, there was never much doubt.

But in aviation, things are different. Visual and perceptual illusions in the cockpit, especially on final approach, can subtly deceive even the most experienced pilot. Such illusory effects are a major cause of accidents even in the most sophisticated aircraft. In fact, it may be that very sophistication that has pointed out the insidious nature of the problem. After all, when visual illusions contribute to accidents in well-equipped aircraft at major airports, it serves to underline the universal nature of the problem.

In late 1973, an airline DC-9 crashed short of the runway after making a coupled and stabilized approach down to DH. Approach and runway lights had been sighted at about 700 feet, or nearly two miles from the threshold. The captain stated that when he released the autopilot at the middle marker, the airplane was in trim, with both flight director and raw data ILS displays centered. As he looked out through the windshield, the runway appeared normal to him for about five seconds, and then "flattened out" in "the blink of an eye." Despite his immediate reaction, the airplane struck the approach lights and crashed 450 feet past the runway threshold.

Ground observers testified that heavy rain was falling during the approach, and the NTSB (National Transportation Safety Board) determined that the probable cause of this accident was "that the pilot did not recognize the need to correct an excessive rate of descent after the aircraft had passed decision height. This occurred despite two verbal reports of increasing sink rate by the first officer. The captain disregarded the reports of the first officer *possibly because of the influence of a visual illusion* caused by the refraction of light through the heavy rain on the windshield."

After a near-perfect approach in difficult weather, this conscientious but unfortunate crew may have fallen prey to one type of visual illusion. There are many:

Water refraction. The most obvious difficulty created by rain and water on the windshield is reduced forward visibility. A more insidious problem is the refraction or bending of light as it passes through that concentration of liquid. The refraction error will cause you to perceive the horizon below where it actually is. The total error can be as much as five degrees, so that at one mile, for instance, the runway or lights would appear to be several hundred feet lower than they really are. Bear in mind that the glideslope is only about three degrees to begin with, so that a five-degree illusion could be fatal. When rain is falling through clear air, the dangers of refraction are exacerbated because all other visual cues remain well defined.

Diffusion of light. Rain or any visibility restriction can substantially affect your judgment of distance to the approach and runway lights by diffusing their glow and causing them to appear less intense. This attenuation of perceived light will suggest that the lights are farther away than they actually are. You will imagine yourself to be high due to the lack of contrast and shadow. Occasionally rainfall will cause lights to appear larger so that the pilot believes himself to be closer than he actually is. In both cases, the natural response is to seek an altitude compatible with the perceived runway elevation.

Slope. When the runway slopes up, the normal glidepath will appear to be too steep, and your eye will lobby for a lower approach. When the runway slopes away from the approach end, the glidepath will appear to be too shallow.

The relative slope of terrain in the approach zone will have an opposite effect. For instance, flying into an airport on top of a hill with upsloping approach zones can create the illusion of being too high.

Runway dimensions. Relative runway dimensions will affect the senses during the visual part of the approach because we all naturally relate to a learned norm. Extra-wide runways will appear to be shorter because we carry a preconceived notion about length-to-width proportions. Narrow runways may give the impression of excess length for the same reason.

Runway contrast. Lack of color contrast seriously degrades depth perception. Concrete runways in the desert, black runways at night and snow-covered runways will challenge your altitude perception to the limit, making you feel farther away and higher than you are. Very bright lights, or bold colors will have an opposite effect.

False horizon. This is a really insidious illusion that occurs on dark nights when ambient surface lighting is not visible. When a steady light suddenly does come into view, it may dominate the pilot's perception with an overwhelming sensation that the light is above the horizon, and that the aircraft should be pitched over to establish an attitude compatible with that false impression. Many pilots have reported an uncomfortable climbing sensation even though they continued to maintain level flight.

Perceptual illusions are compounded by the fact that simple vision is always biased by other sensory and psychological cues. The balance mechanism of the inner ear can grossly alter or prejudice any and all sensory input. When we are fatigued, we are especially prone to overlook or ignore indications that may demand a response.

There is, in addition, one unusual visual sensation that can cause fatigue . . . or worse.

The Flicker
Effect

8

Strobelights. Rotor blades. Ragged clouds. Slow-turning propellers. All have one thing in common. They produce a rhythmic or wavering light pattern—a flicker—that can induce serious physical distress in both pilots and passengers.

After a rather lengthy flight, the pilot of a single-engine prop airplane made a normal landing. Following the rollout, his airplane remained motionless in the center of the runway with its propeller revolving slowly. The pilot was found unconscious. Rays of the setting sun flashing through the propeller blades had induced a strobe-effect seizure.

Five minutes after departure, the passenger of a heavy twin-engine helicopter suffered what appeared to be an epileptic seizure. He first went into convulsions and then stopped breathing. Crew-members applied cardiopulmonary resuscitation and the patient resumed breathing on his own. He regained consciousness after several minutes but was seriously disoriented for another half hour or longer. The passenger eventually reported that the sun shining through the rotor blades had been mildly discomforting and closing his eyes did not eliminate the stroboscopic effect.

Doctors diagnosed his collapse as a "strobe-effect-induced grand mal seizure," even though this individual had no history of such seizures and no abnormal tendency toward them.

A pilot who was working on the ramp directing a prop-driven transport to its parking space collapsed in front of the approaching airplane. After recovery at the local hospital the victim mentioned that he had been hypnotized by the flickering effect of the setting sun shining through the spinning props.

All three were victims of a form of electrical interference.

The human brain produces periodic, or rhythmic, electrical currents that can be measured and recorded. The electroencephalograph (EEG) is a device designed to do just that. One of the predominant electrical signals produced by the brain, the alpha wave, is normally between eight and 12 hertz, or cycles per second, in frequency. A light flickering at between one and 40 cycles per second can upset the alpha wave and cause serious physical problems.

The phenomenon has been called "flicker vertigo," but that label is misleading. Vertigo, or disordered equilibrium, is just one possible result of flicker's influence on the alpha wave. Accordingly, the term "flicker-induced seizure" is now accepted by the medical profession because it is more appropriate and more accurate. Essentially, there are two broad effects.

In its milder form, flicker-induced interference with the alpha wave is more annoying than debilitating. It may be limited to a mild mental and physical irritation, but it also can result in fatigue, dizziness, drowsiness, nausea to the point of vomiting, and disorientation. These are, by far, the most common physical effects of flicker. They can render a pilot unfit to fly, if not incapacitated, and they can produce marked discomfort in passengers.

The less common but more serious effect of flickering light is a grand mal seizure, the very same type of collapse that afflicts victims of severe epilepsy. The grand mal seizure results from what is called photic-driving, which pushes the alpha wave into self-acceleration. Fortunately, very few people are susceptible to the photic-driving, but when it happens, the grand mal seizure is dramatic.

Initially the victim may complain of a generalized strange feeling, but there may be no warning at all. When the seizure begins, the victim may cry out. If he is standing, he will fall to the floor unconscious. His muscles will become tense, and the whole body will be rigid.

Breathing may stop temporarily, which will cause the victim's face to turn blue. The muscles will then begin to jerk spasmodically as breathing resumes. Breathing may be labored if the tongue has fallen back, obstructing the airway.

The victim may bite his tongue, foam at the mouth, void his bladder and bowels, or all of the above.

Although the seizure itself usually is over in a few minutes, the victim may be unconscious or semiconscious for some time afterward.

Clearly the victim of a grand mal seizure needs immediate first aid and a great deal of caring support as he recovers. The following steps should be taken if one of your passengers or fellow crew-members suffers a flicker-induced seizure.

• If possible, place a gag between the patient's teeth to prevent him from biting his tongue. Don't try to force anything between clenched teeth, however, and *never* use your fingers as a gag. A tongue depressor wrapped in a large handkerchief will suffice.

• Loosen the patient's clothing.

• Make sure that the patient resumes breathing. Check for an open airway, and use CPR if necessary (although it rarely is).

• Restrain the patient enough to prevent injury during the jerking stage, but allow all possible freedom of movement since unnecessary restraint can cause injury.

• After the seizure, keep the patient warm and quiet. Turn his head to one side to prevent choking if he vomits.

• Any seizure victim should be evaluated by competent medical personnel as soon afterward as possible.

Flicker-induced incapacitations are rare, but when they do occur they can range from mild discomfort to total collapse. Fortunately, there are some positive actions you can take to minimize the exposure and the risk.

Essentially you must be aware of situations that will involve some intense, rhythmic flickering of light in a frequency range of one to 40 hertz. In aviation, that situation arises from several unrelated sources, but some of the most common are:

• Sunlight passing through helicopter rotor blades.

• Sunrise or sunset viewed through an idling prop either during slow descent or while on the ground.

• The flashing effect of sunlight when you are flying through ragged cloud tops.

• The reflections of anti-collision lights or navigation lights when you are flying at night or in clouds.

When you recognize the right circumstances for flicker-induced problems, take whatever action is necessary to eliminate the flicker. Turn off the lights. Change altitude or heading. Close the cabin window curtains to protect passengers. Increase propeller rpm.

You cannot eliminate the threat by force of mind. Flicker-induced problems are physiological phenomena, and if you are susceptible, they will affect you. You must stop the flickering in that critical frequency range of one to 40 hertz to prevent the possible ill effects.

THE FLICKER EFFECT

At the bottom line, the airborne remedy for all sensory illusions and distractions is simple:

Always use precision approach facilities when available. Give a high priority to ILS or VASI glidepath information. Barring that, compute your own glideslope by using any available distance information and figuring 300 feet of altitude for each mile of distance. It's really the same old story—stabilized approaches, crew coordination and attention to cockpit instrumentation, particularly rate of descent.

Biorhythms

9

Human limitations of sight and fatigue and age are a fact of life. But suppose those limits were more predictable. Wouldn't we be better able to cope?

In the past several years, the theory of biorhythms has been injected into the aviation environment as a potential means of enhancing safety. Several authors have extolled the virtues of this speculative scheme, and some government agencies and air carriers have assessed its potential in maintenance shops and cockpits. It's a fascinating and interesting subject. But, the essential question is, how well do biorhythms correlate with aviation safety?

The history of biorhythms begins at the turn of the century when Dr. Herman Swoboda, a Swiss psychologist, discovered that musical melodies and ideas were sometimes repeated in the mind on a 23- or a 28-day cycle. At the same time, Wilhelm Fliess, a German physician, was pursuing his theory that every individual possesses some elements of bisexuality.

Fliess believed that a 23-day masculine rhythm affected man's physical condition while a 28-day feminine rhythm influenced female emotions and sensitivities. Later, in the 1920s, Alfred Teltscher, a Swiss teacher, reported that he seemed to observe a 33-day intellectual cycle in the performance of his students.

Those three theoretical cycles, 23-day physical, 28-day emotional and 33-day intellectual, are the foundation of biorhythm logic. Believers maintain that each of the three cycles begins at the very moment of your birth and continues in sine wave fashion with absolute constancy and precision until your death. The first (positive) half of each cycle is regarded as a time of strength, while the second (negative) half is a period of weakness as the system recharges itself. So-called critical days occur when one or more of

37

the cycles crosses the zero line from positive to negative or vice versa.

By way of example, the 28-day sensitivity cycle would include 14 up days followed by 14 downers. At the instant of birth, the curve starts rising until it peaks 168 hours (seven days) later and then begins to fall. At 336 hours (14 days), the cycle switches to negative during a critical day. Now the curve bottoms at 504 hours (21 days) and starts back up until it crosses back to positive at 672 hours (28 days) which is another critical day.

Biorhythm practice involves calculating your own personal cycle by using a special calculator or computing device programmed with your birth date and the day in question. Then, supposedly, when you know where you are physically, emotionally and intellectually, you can plan your schedule accordingly.

Devotees are quick to say that critical days are not inherently dangerous, but they are days during which a person should be particularly careful. Double or triple critical days, when two or three cycles are switching in phase, are deemed to be particularly fraught with a sort of negative potential. Not dangerous, you understand, just critical.

It is possible that biorhythms play some part in human actions and reactions, but consider the following points when you assess the direct safety implications of this very moot theory:

Critical days are not uncommon. Each of the three cycles, fluctuating in sine-wave fashion, crosses the zero point twice each full cycle. Therefore, two of the 23 physical days are critical, as are two of the 28 emotional days and two of the 33 intellectual ones. Most people define the critical period as 24 hours, which means that over 20 percent of all the days in your life are critical, when adjustment is made for multiple critical days. Some investigators have used a 48- or 72-hour critical period to establish a relationship between critical days and accidents so that the percentage of critical days quickly becomes a majority.

Accident occurrence has not been shown to correlate with critical days. In 1975, four investigators analyzed the biorhythm accident probability of 8,625 cases. Data were tested by the chi-square statistical test, which established a relationship between *expected* results—in this case, ignoring any possible biorhythmic influence —and *observed* results. If the biorhythm theory is correct, there should have been a higher incidence of accidents on critical days. Some previous reports have claimed that a person is three to eight times more susceptible to accidents during a critical period. It might also be reasonable to expect more accidents on days that fall in the negative half of the cycle.

When all the math was finished, the study of those 8,625 cases failed to demonstrate that aircraft accidents respond to the biorhythm theory. The total percentage of accidents that occurred on critical days amounted to 20.78 compared with an expected ratio of 20.36. That deviation is not statistically significant.

Coincidence is not to be confused with establishing a cause and effect relationship. Biorhythm advocates like to use selected experiences from the lives of famous people (like Winston Churchill or Arnold Palmer) to document their theories. Coincidence, while interesting, is not documented proof. The simple law of averages would yield a multitude of circumstantial evidence.

Cycles per se are not in doubt. Few would deny that life is a series of ups and downs. We all have good and bad days physically, intellectually and emotionally. There is, however, reasonable doubt that such cycles occur with absolute and inviolable precision, unaffected by sickness, accident or trauma. Anything less than perfect stability in those three cycles would preclude useful predictions.

The power of suggestion is very strong. The poor guy who believes in biorhythms cannot help being influenced by his chart for the day and then read into it his own fears and expectations. Autosuggestion is a classic and common psychological phenomenon, applicable to biorhythms and many other hare-brained attempts at pseudo-psychology.

Obviously, there are sure to be some good days and some bad ones in everyone's life. If we, however, just accept flying as a "critical pastime," and exercise appropriate care, safety will naturally follow. Anything else is too likely to become an excuse for poor performance. Let's face it. If we didn't have human limitations we wouldn't need airplanes. Conversely, the limitations that we do have are often magnified in the aviation environment. I guess I have known and flown with 1,000 pilots at least and I don't hesitate to make a sweeping generalization: most of us are far more conscious of technical and mechanical limitations than we are of the flesh-and-blood variety. When you think of it, that mind-set in itself is just one more human limitation.

SURVIVING THE SYSTEM

II

They design a magnificent new airliner and introduce it into service. It is larger and faster and flies higher and quieter than anything else. They are very proud, but there is a tragic flaw. When the fuel-dumping system is used on those rare occasions that require it, the vapors from the fuel are ingested into the cabin's combustion heater, causing a massive explosion. It happens three times, with great loss of life, before the problem is identified and corrected.

They write an involved set of regulations governing the planning and execution of takeoffs in large airplanes. The regulations are detailed and precise, but they address only one possible problem, and that one inadequately. They are a model of linguistic precision, but airplanes routinely crash on takeoff because the substance of the regulations is flawed.

They pass a law about how to fly more quietly so that airplanes will not bother people on the ground. They appear to be champions of the people when, in fact, they have jeopardized countless lives for the sake of a few disputed decibels.

They pressure the pilots to fly with too little sleep, fuel, runway, visibility, instruments, engines, equipment and maintenance. Someone else could do it, why not you?

Surviving the system is largely a matter of surviving "them." They are an elusive group which dogs the steps of every pilot, but they can be identified in two ways. First, they are nonflyers. They do not commit their lives and careers to the potential hazards of flight. Second, they have some responsibility for the process of flight, which makes their decisions and actions of particular interest to the pilot.

They are management, air traffic controllers, mechanics, dispatchers, meteorologists, cargo handlers, politicians and aircraft designers. They mean well for the most part, and they try hard, but they do not sit in the hot seat.

I have enjoyed the support of thousands of other people during my aviation career, and I truly appreciate their effort and hard work. I also know that, as the pilot in command, I am the very last line of defense. I am at the cutting edge of success or failure. It is my job to survive the system.

ATC Radar Problems

10

Just as we have personal, human limitations, so the air-traffic-control system has limitations of its own.

Several years ago, on departure from Los Angeles International, on a scheduled airline trip, we raised the landing gear, but it would not fully retract despite our best efforts at troubleshooting. Unable to continue the flight and reluctant to risk unnecessary damage by another retraction attempt, we requested a tower flyby for a visual check. Those alert controllers were quick to notice that the four-wheel, left main truck was angled off the horizontal enough to prevent the gear door from closing. The broken "bogie leveler" causing the problem did not prevent a safe return for repairs.

Pilots often take for granted that air-traffic controllers understand the implications of engine shutdown, hydraulic failure or pressurization problems. But too often that professional overlap is lacking from the cockpit side, although pilots would benefit from a knowledge of controllers' problems and the air-traffic-control system structure.

Controllers' problems begin with the complex equipment on which the whole system is predicated.

A few years ago, the entire ATC system changed from the older broadband radar to a highly sophisticated system called narrowband or Radar Data Processing (RDP). In the RDP system, raw broadband information is digitized at the antenna site and fed, along with other data, into a computer. The computer then generates one comprehensive picture on a TV-type screen. Controllers do not see the actual reflection of your plane or transponder. They watch stylized television images of computer calculations which they refer to as narrowband. The implications are enormous, because the computer itself represents a significant new element capable of misrepresenting, losing or garbling the vital information that allows for adequate traffic separation.

Initially, RDP was the source of some rather serious system errors as controllers and technicians adjusted to the new equipment and new techniques. Even now, there is an inventory of very real failure in the RDP system. Let's look at some.

Coasting. High-speed airplanes can literally outrace the computer's tracking ability. When that happens, the great iron brain just gives up and reverts to a prediction memory to project where the airplane is likely to go based on the previous speed and course information. That dead-reckoning type of prediction is called "coasting."

Target swapping. Remember that the controller does not see actual radar images of the aircraft in his sector. He watches a TV picture, constructed by the computer, including data blocks that follow each aircraft having a discrete (4096) transponder. Target swapping is a computer malfunction in which the data block for one aircraft is mistakenly assigned another.

When RDP first went on line, there were numerous reports of data blocks being assigned to incorrect targets. Much of the original problem has been eliminated although target swapping is still a potential problem with nondiscrete transponders.

Ringaround. Within 15 miles of the radar antenna, reception can be greatly distorted. One result can be that a single aircraft appears as a number of separate targets in a straight line, each with a partial data block. The controller is then forced to speculate about which return might be the actual target while attempting to assemble and assign the full data block.

Much of the early ringaround problem has been eliminated with the use of absorbent material around the antenna.

Jitter. Jitter is a freaky side effect of the RDP concept. In order to envision it you need to understand better the makeup of the scope picture itself.

That picture is actually divided into a number of boxes, each representing a 16-square-mile area. In each box, the aircraft are read by a different radar component. As a target passes from one box to another, it passes through a so-called slip box seam between those two areas of radar coverage. Right at that moment two different radars are actually contending with one another to place the target in their box. During that transition the target will appear to crab sideways, with a back-and-forth motion as much as

44

four or five miles on either side. When a series of airplanes are involved, that jitter can be distracting and confusing.

Primary cell error. This problem is a close kin to jitter. When those two separate radar components contend for the same target at the slip box seam, they each may register the plane as a different target within their respective blocks, resulting in two individual targets.

Primary clutter. Controllers don't like the assorted dots and pulses programmed into their display to indicate obstacles and high terrain. They feel that this primary clutter introduces unnecessary distraction. It may be distracting for them, but as a pilot I rather like the thought that those cumulo-granite are constantly registering their presence with the traffic director.

Primary target. The new narrowband display does not paint nontransponder traffic as well as the old broadband equipment did. Controllers now revert to the standby broadband when there is a discrepancy involving primary targets. When and if the broadband option is decommissioned, even that alternative will not be available.

Weather display. The narrowband (RDP computer display) does not paint weather as well as the broadband (raw radar) so that some controllers complain about traffic management in areas of severe buildups. That objection may lack some credibility, however, in light of ATC's long-standing reluctance to accept responsibility for weather avoidance. When broadband radar was the primary equipment, circular polarization circuits were employed to expressly eliminate weather returns. RDP may actually improve the weather-depiction problem through its unique capability to present stylized information without blanking large areas of the scope.

Some of these difficulties are the normal consequence of any complex system, requiring adjustments on the part of each ATC employee in the same way that pilots must adjust to the constant juggling of airspace. In fact, it's nearly a full-time job to sort out the distinctions between Group II and Stage III, so let's start at square one.

Terminal Control Area 11

In the beginning, all domestic airspace belonged to 20 air-traffic control centers. Where appropriate, segments of that airspace were allocated by letter of agreement to local controlling authorities for uses such as military training and test areas, approach and departure control zones, and TCAs.

In recent years, the airspace system has become even more complex with the addition of three different groups of TCAs, three separate stages of radar service, Terminal Radar Surveillance Areas and hundreds of plain control zones. The terminology alone is enough to confuse a pilot, so let's take a few minutes in this chapter to sort out these distinctly different airspace segments.

Airport traffic patterns have been established at all facilities to designate altitudes and directions for the orderly flow of traffic. Your only responsibility is to adhere to the prescribed pattern at the airport you are using and to see and avoid other traffic.

An airport advisory area is the airspace within five miles of an airport where a control tower is not operating but where a flight service station is. At these locations, the FSS provides advisory services to arriving and departing aircraft but assumes no ATC responsibility. Pilot participation is strongly recommended but not mandatory. Basic traffic pattern rules apply.

Airport traffic areas exist at all airports where a control tower is in operation. In most cases the ATA is a cylinder five miles in radius, centered on the airport and extending from the surface up to, but not including, 3,000 feet agl (above ground level). This airspace is reserved for departures and arrivals of aircraft at the controlled airport, and all other flight is prohibited unless cleared by the tower.

Unless prior arrangements have been made, you must be in radio contact with the control tower. Whenever the control tower shuts down, even overnight, the airport traffic area ceases to exist and basic traffic pattern rules apply.

Control zones provide separation between instrument flight (IFR) and visual flight (VFR) traffic near airports with approved instrument approaches. A control zone is normally a circular area within a five-mile radius of the airport and includes any extensions needed to accommodate instrument departures and approaches; the zone may be enlarged, however, to include more than one airport. It extends from the surface to 14,500 feet asl (above sea level), the base of the Continental Control Area. A control zone simply establishes controlled airspace, and its accompanying higher VFR minimums, where it might otherwise not exist. Many airports have instrument approaches but no control towers and the floor of controlled airspace above them would start at 700 feet, were it not for the control zone.

Below such a floor, a pilot legally could be VFR with only one statute mile visibility, if he stayed clear of clouds.

Stage I radar service provides VFR aircraft with radar traffic advisories and limited vectoring if the controller's workload allows. Officially, Stage I service is available at nine terminal radar facilities but in reality it is available anywhere there is radar coverage.

Stage II radar adds another level of service for VFR traffic. In addition to traffic advisories, Stage II provides full-time sequencing to VFR aircraft landing at the primary airport. Fifty-eight locations offer Stage II service.

Stage III radar service provides traffic advisories, sequencing and separation between all participating aircraft operating in an established Terminal Radar Service Area or Terminal Control Area (where participation is required). That introduces two more airspace segments for us to define.

Ironically, Stage II service is available at San Diego—where a well-known disastrous midair collision occurred—and Stage III is also in use there above 4,000 feet.

Participation in all three stages of radar service, which is provided as part of the FAA's Terminal Radar Program, is voluntary, but the agency strongly urges that all pilots make use of them.

Terminal Radar Service Areas (TRSAs) were established by the FAA solely to define the airspace in which Stage III service is provided. If Stage III is available, you are in a TRSA. If you are in a TRSA, Stage III is available.

Terminal Control Areas (TCAs) consist of controlled airspace with special operating rules and pilot and equipment requirements. Terminal Control Areas were established by the FAA to minimize the risk of midair collisions at heavily congested terminals. In theory there are three types of TCAs; in practice there are only two.

Operations in Group I TCAs require:

- ATC authorization.
- A private pilot certificate or better if a takeoff or landing is to be made.
- An operating VOR receiver.
- An operating two-way radio with appropriate frequencies.
- A 4096-code transponder.
- An encoding altimeter.

In Group II TCAs the encoding altimeter and private pilot certificate are not required.

Group III TCAs exist in concept but have never been established. Group IIIs would require basically a "talk" or "squawk" capability but the FAA has shelved all plans for Group IIIs.

Airspace is becoming ever more complex and terminology alone is enough to discourage even the most serious professional. And that terminology extends to the approach clearances that may be issued by arrival controllers.

An old friend had a stock answer for the tower when they cleared him to land. "Ooh," he'd say, keying the mike, "that's the part I hate." Obie didn't really hate landings at all. He was a fine pilot. But if he were still flying, he might accept them with less enthusiasm in light of the tangled options that govern that final approach.

Approach Options

12

We have had to cope with an explosion of new regulations in the past ten years. Operating rules for controlled and uncontrolled air traffic have grown in number and complexity so that pilots often are faced with a confusing assortment of operational choices. Nowhere is this so true as in the landing approach.

Arriving pilots must cope with the legal distinctions between groups of terminal control areas, stages of radar service, and visual, contact, IFR and special VFR approaches. Most air traffic controllers understand the approach alternatives very well. Ironically, pilots, who bear the final responsibility for the safety of flight, often are confused and sometimes intimidated by the rapid-fire sequence of events and options in the terminal area.

Basically, the pilot of an IFR flight will receive one of four possible clearances in the terminal area: he may complete his IFR flight plan without modification; he may cancel IFR and continue VFR; or he may accept a visual or a contact approach. In each case, his authority and responsibility are different.

The pilot of a VFR flight has no real options. He must find the airport and land visually, providing his own traffic separation. He can, in some instances, request a special VFR clearance if the ceiling is below 1,000 feet and the visibility is less than three miles, although he must always operate clear of all clouds. At some terminals he may request radar advisories, sequencing and separation, but more on that later.

In most cases an IFR trip will terminate with a routine instrument approach and landing. When visibility is one mile or better, you may request one of several options. If weather conditions are VFR, you can cancel the IFR flight plan and navigate visually. Such a choice often saves time but it also relieves ATC of any

51

responsibility for traffic separation. Not too bad a tradeoff at smaller airports but often a poor decision at busier terminals. Of course, if the weather is VFR, the pilot has a responsibility to maintain separation from all uncontrolled traffic whether or not he is operating under an IFR flight plan.

A far more common option is the visual-approach clearance, which may be issued by ATC to arriving IFR traffic when the ceiling is at least 500 feet above minimum vectoring altitude and the visibility is at least three miles. The visual approach is a tactic for reducing controller workload and integrating IFR and VFR arrivals. To the pilot, it offers some compromise between safety and efficiency.

The FAA defines a visual approach as: "An approach wherein an aircraft on an IFR flight plan, operating in VFR conditions and having received an air traffic control authorization, may deviate from the prescribed instrument approach procedure and proceed to the airport of destination by visual reference to the surface." It should add "or by visual reference to a preceding aircraft."

When ceiling and visibility allow, ATC can vector you to the airport area, instead of the final approach course, and then clear you for "the visual" when you have the airport or the preceding traffic in sight. When following another IFR flight, you can be cleared for "the visual" if you have only the airplane in sight and the tower is informed of your position by the approach controller. You need not have the airport in sight.

The visual approach is invariably easier on air traffic controllers than an IFR procedure is. It can be of dubious value, however, and you are not required to accept it. Although the visibility is three miles at the airport, it may be markedly less at your position five or ten miles away. In fact, it is not uncommon to be cleared for that visual at some substantial distance from the airport on the strength of having reported the preceding airplane in sight. Many pilots will accept a clearance to follow some barely distinguishable traffic through 15 miles of murk to the runway.

Accepting a visual when the field is in sight 40 miles away also can create a problem. The navigation is simple, assuming you have the correct field in sight, but you will be giving up ATC's help with traffic separation as you slice through a deep altitude structure under conditions conducive to heavy VFR operations. If you accept that 40-mile visual, ask for radar traffic advisories or actual separation while keeping a good watch yourself for other traffic.

Charted visual approaches such as those found at SFO, DCA and LGA are even worse. They have no legal standing despite the official appearance of the approach plate. They are local procedures designed by gentlemen's agreements and do not conform to FAR (Federal Air Regulations) Part 97 requirements.

The visual approach is a hybrid; it allows IFR traffic to operate by visual flight rules, with reduced separation criteria; it allows instrument traffic to follow one another in a flimsy daisy chain through an area of marginal visibility. And it seriously blurs the lines of authority and responsibility between pilot and controller.

The other less common option is the contact approach, which the FAA defines as: "An approach wherein an aircraft on an IFR flight plan, operating clear of clouds with at least one-mile flight visibility and having received an air traffic control authorization, may deviate from the prescribed instrument approach procedure and proceed to the airport of destination by visual reference to the surface."

The contact approach is of limited value for several reasons. You must have at least one-mile visibility and be clear of all clouds; you must have ATC separation from all other aircraft, and you— as pilot in command—must specifically request a contact approach. The contact approach can be useful when making a circling approach, although neither pilot nor controller really benefits from this relatively rare clearance.

One more note on IFR terminations. If you operate under FAR Part 91, you—as pilot in command—always have the authority to begin an approach, regardless of the reported weather. Then, if things look bad at the DH or MDA (Decision Height or Minimum Descent Altitude), you simply fly the pull-up procedure and make some other plans. If you switch, or alternate, to an FAR 135.2 operation, you must satisfy FAR 121.651, which requires that reported weather be above the approach minimums before you may begin the approach. You may continue the approach to the prescribed limit if the visibility falls to below the minimum after you are inside the outer marker on an ILS, have reached the MDA on a nonprecision approach or have been switched to the final controller on a PAR.

Arrival procedures are somewhat more complex than they used to be, but the root problem is command authority. FAR 91.3 delegates final authority for aircraft operations to the pilot in command. Still, that authority cannot be exercised properly without a working knowledge of the airspace environment. In recent years the basic concept of the control zone has been expanded and modified with the addition of TCAs, the Terminal Radar Program

53

for VFR aircraft and Terminal Radar Service Areas (TRSAs). Here's a brief outline:

A control zone creates controlled airspace in a defined area around an airport with an approved instrument approach. To increase safety at high-density airports, the FAA added Terminal Control Areas at a number of locations. To operate in a TCA, you must meet certain equipment and pilot certification requirements and ask for and receive an ATC clearance, even under visual flight rules.

The agency also established a voluntary Terminal Radar Program for VFR pilots who want to take advantage of the ATC radar coverage available in many areas. Various levels of traffic advisory, vectoring and sequencing services are provided in these areas depending on which of three stages is available. A TRSA simply designates an area with Stage III service. Pilots who turn down these services are forfeiting valuable assistance that can improve safety and aid traffic flow for all operators. A pilot should think seriously before declining such assistance. Pilots also should be aware that since participation in these three stages of radar service is not mandatory they must keep a sharp eye out for traffic and not rely solely on ATC for adequate separation.

There are several inherent limitations to approach control and tower control in terminal airspace. Air traffic controllers cannot and should not be expected to bear all the responsibility for traffic and terrain separation. You must do your part in the cockpit:

• Always pre-brief for the approach in use, prior to entering the terminal area. Frequencies, altitudes, obstructions and missed approach procedures should be reviewed and noted well in advance.

• Never allow radar control to take the place of your cockpit navigation. Use every possible navigational aid to follow your flight progress and question the controller if altitudes or positions are not appropriate. Remember that your weather radar is an excellent navigational device and that any facility that can give you a bearing, such as beacons or broadcast stations, can be helpful. If you do become disoriented, however, simply request your position from the controller. Above all, be constantly aware of your position and track and keep in mind approach MOCAs.

• VFR pilots should not presume that radar advisories in a TCA or TRSA are a substitute for visual separation. Radar service to VFR traffic never absolves the pilot from his basic responsibilities to maintain visual navigation and separation.

APPROACH OPTIONS

• Never accept a visual approach clearance casually. Both the NTSB and the FAA recently have questioned the "see and be seen" concept. Our eyes may not be as useful as we would like to think, and they are *useless* if you don't use them.

• Use every light on your airplane in any terminal area. One study indicated that landing and taxi lights make an airplane more conspicuous than all its strobes and beacons combined. Use every light for each departure and arrival.

• Visit an FAA tower and approach-control facility to see that end of the operation firsthand. Pilots and controllers make this system work and they need to share each other's problems.

Now, if you think that approach options are confusing, consider the business of landing minimums.

So What's an RVR?

13

As you know, the basic concept used by the FAA is that visibility is the sole operating minimum for takeoff and landing, although there are four separate methods of reporting that visibility. RVR, RVV, Sector Visibility and Prevailing Visibility, in that order of priority, could be the controlling report for your departure or arrival, depending on circumstances. Ceiling minimums are no longer used except for alternate airports and a few takeoff minimums under FAR Parts 121, 123, 129 and 135.

Runway Visual Range (RVR) is always the controlling element when reported. In that case, the other three reports are advisory only, and you must predicate your actions solely on the tower-reported RVR.

RVR readings are derived from electronic measurements made with a device called a transmissometer. This unit is actually comprised of two separate elements, a transmitter and a receiver, spaced 250 feet apart, abeam the touchdown zone of an ILS runway. The transmitter is a precisely calibrated light source. The receiver effectively measures how much of that light penetrates the atmosphere in the intervening space. Results are averaged over 60-second segments and transmitted as impulses spaced one minute apart. In National Weather Service offices or Flight Service Stations, the impulses are received and recorded continuously on paper graphs. These recordings, no more than raw data, must then be modified into tables that take into account prevailing daylight, if any, the runway light setting in use and anything else that might alter ambient lighting conditions in the vicinity of the transmissometer. The resultant RVR is reported at the end of NWS sequence reports and to operations offices on the airport.

Airport control towers and approach control rooms receive RVR in digitalized form. The raw impulses from the transmitter/receiver receive a direct and simple digital readout. These are up-dated only once each minute so the tower RVR cannot change more often.

Each airport with RVR equipment has one designated runway for sequence reporting purposes. Chicago O'Hare, for instance, has five RVR runways, but 14R is the designated installation. There-fore, the RVR for 14R will appear on the Aviation Sequence Report whenever the prevailing visibility is one mile or less, or when the highest RVR reading on 14R is 6,000 feet or less. The RVR will appear at the end of the sequence as a range of values representing the highest and lowest RVR recorded during the ten minutes pre-ceding the report.

Actual airport operations are governed by the digitalized read-ings in the tower and approach control. These instruments nor-mally read from 1,000 to 6,000 feet in 100-foot increments. Some read down to 600 feet. The raw numbers are modified by use of a "+" or "−" to indicate more subtle differences. Thus, 2700+ indi-cates something over 2700 RVR but less than 2800 RVR; 2700− shows something less than 2700 but better than 2600 RVR. When minimums are 2400 feet, for example, a reading of 2400− is below minimums. The maximum scale, then, is from 1000− (or 600− in a few cases) to 6000+. RVR cannot measure conditions above or below these limits.

All Category II runways and a few Category I runways are equipped with a second transmissometer at the rollout end. Regu-lations concerning the use of rollout RVR have changed several times since the inception of Category II, but that situation has not apparently stabilized. When RVR at the approach end is below 1600, rollout RVR is controlling for takeoffs (for some operations) and advisory for landings, in the following manner:

• For takeoff, RVR at both ends is controlling for FAR Part 121, 123, 129 and 135 operations.

• For landing, rollout RVR is always advisory. During Cate-gory III operations, rollout RVR is provided when it is less than touchdown RVR. For basic ILS operations, RVR for the rollout end is provided if it is less than 2,000 feet and less than touchdown RVR. In all cases, rollout RVR is purely informational for landing.

RVR readings for other than the single designated runway at each airport are only available from the local controlling facility.

Naturally, RVR minimums require no ceiling report, but remem-ber one other fact: Runway Visual Range is only an electronic measure of the horizontal visibility for 250 feet of the touchdown zone. It is very local in nature and may not correspond at all to the nearly three fourths of a mile slant-range visibility you will require for the last 200 vertical feet of your landing approach. Nevertheless, when RVR is reported, it must be used as the controlling information, regardless of any other visibility reports. Where there is no RVR, then RVV, Sector Visibility and Prevailing Visibility, in that order, are controlling.

Runway Visibility Value (RVV) is measured by equipment similar to that used for RVR except that measurements are in miles and/or fractions thereof. RVV is the controlling report for that runway for takeoffs and landings. Prevailing Visibility is controlling for all other runways.

Sector Visibility, or Quadrant Visibility, is sometimes reported when visibility in one direction is markedly different. This Sector Visibility is then controlling in that direction while Prevailing Visibility controls everything else. If, for example, the hourly sequence or tower controller reports VSBY E 1¼ MILES, that visibility is controlling for aircraft departing to the east or arriving from the east.

Prevailing Visibility is controlling for everyone when no other reports are given. At most airports, the Prevailing Visibility as reported in the NWS sequence report is the legal visibility when it is four miles or greater. When visibility is below four miles, controllers have the responsibility to establish Prevailing Visibility.

Watch those visibility reports. Sometimes you need a lawyer to determine the legal minimums.

Missed Approaches

14

Understanding your legal weather minimums for an approach and landing can be a real challenge. Actually coping with those minimums when they exist is something entirely different. I knew one guy who busted a checkride over that issue in a DC-6 simulator.

The legal approach minimums for that simulated ILS approach to Runway 26L at Denver were 200-foot ceiling and ½-mile visibility in those simpler days of yore. This particular guy did a creditable job of maneuvering the DC-6 down the glideslope to 200 feet when the checkpilot called for a missed approach. With a quick 90-degree to 270-degree turn this pilot reversed course and climbed right back up the ILS course to hold at the outer marker. When questioned about this illegal, unorthodox, crackpot procedure, he had an immediate answer: "I didn't hit nothin' comin' in, so I figured I wouldn't hit nothin' goin' back out the same way."

The logic sounds good but there is more to the missed approach than homespun logic. In fact, I have a confession to make, myself.

In the last several years I have seldom paid proper attention to missed approach procedures (except when the weather was so bad I fully expected to miss even before beginning the approach). Fortunately, there have been precious few misses, so my lack of preparation has not bitten me very often.

Maybe success is responsible for the problem. Experience reinforces our expectation of making a successful landing, so we lose incentive to properly prepare for a one-in-a-thousand possibility.

Actually, it's a matter of priorities. A simple review of FAR 91.117, "Limitations On Use of Instrument Approach Procedures," establishes those priorities. Notice the wording of paragraph (b):

61

Descent below MDA or DH. No person may operate an aircraft below the prescribed minimum descent altitude or continue an approach below the prescribed minimum descent altitude or continue an approach below the decision height unless—

(1) The aircraft is in a position from which a normal approach to the runway of intended landing can be made; and

(2) The approach threshold of that runway, or approach lights or other markings identifiable with the approach end of that runway, are clearly visible to the pilot.

If, upon arrival at the missed approach point or decision height, or at any time thereafter, any of the above requirements are not met, the pilot shall immediately execute the appropriate missed approach procedure.

Notice the construction of this regulation. It makes the MDA or DH the point at which a missed approach will be made *unless* you happen to see the runway or approach lights well enough to continue. The concept is valid, although it conflicts with pilots' and controllers' experience in a regrettable way. Instrument approaches have such a high incidence of success that we tend to reverse the logic. In real life, the MDA or DH becomes a point from which the landing *will* be made, not the reverse as required by 91.117(b).

I believe that I am not alone in my nonchalance toward missed approaches. There are at least three reasons for this attitude of apparent indifference to the only alternative a pilot has when a safe landing is not possible:

• The IFR environment is so totally controlled we take it for granted that the controller will lead us by the hand in the event of a missed approach. If there were no radio traffic and we really felt isolated in the cockpit, I think we would be much more inclined to plan ahead.

• It is a cop-out to assume that the missed approach procedure will be so simple that a review is unnecessary. We often find, however, that missed approach procedures are more complicated than the approach itself. Look at the ILS Runway 19L approach at San Francisco, for instance.

In order to complete successfully the ILS approach and landing to SFO's Runway 19L, you really need to know only four things: the localizer frequency (108.9), the localizer course (191°), the intercept altitude (1800 feet) and the decision height (285 feet). The missed approach procedure, on the other hand, requires a knowledge of at least nine facts: two altitudes (410 feet and 1900 feet), a programmed left turn, the ISFO localizer frequency

(109.5), the inbound course (281°), the Oakland VOR frequency (116.8), the intersection radial (173°), the left-turn hold and the appropriate entry to that holding pattern (tear drop).

Again, because it so seldom happens, we develop a strong assumption that the missed approach will never be needed; hence the subtle feeling that a review of it is unnecessary. This mental conditioning, which researchers call "expectancy" or "set," is defined as an anticipatory belief or desire. Mixed up in the whole thing is the fact that decisions often are based on how we would like circumstances to be rather than on reality itself.

In its most general form, expectancy works like this:

• We learn a basic skill like flying in an academic, student-teacher environment.

• As we develop that skill, we begin to attach subconscious meanings to the stimuli associated with it. Picture the numeral "3" written twice, once over the other with a line beneath as in an arithmetic problem. If I ask for an answer, you will respond by saying either six, nine or zero, depending on whether you are "set" to add, multiply or subtract. In an airplane we are set to land.

• Once a set or anticipation is developed, we become influenced by it instead of the reality of the situation. The reality of that arithmetic problem is not established until the appropriate sign is added. Likewise, the reality of an instrument approach may not be established until the MDA or DH is reached.

• Finally, our mental set eliminates the viable alternatives by focusing our thoughts on just one possibility such as addition or, in the case of an approach, landing.

The solution to mind set is to recognize its potential and then control the false anticipations. In terms of missed approaches, that solution is a careful review *and* discussion before the approach begins.

Years ago I learned a simple checklist for instrument approaches; I call it WIMTIM:

• Weather checked.
• Instruments and radio set.
• Minimums.
• Time inbound (for nonprecision approaches).
• Instruct other crewmembers about your needs and intentions.
• Missed approach procedure review.

It's time to put the last M back into WIMTIM.

Mid-Air Collisions

15

The air traffic control system is survivable. Thousands of safe aircraft operations prove that every day.

The air traffic control system is also far from perfect. Hundreds of grisly accidents prove that every year.

One of the watershed accidents in recent years was the mid-air collision of a light single with a B-727 near Lindbergh Field in San Diego. There has been lots of finger-pointing, and all of us in aviation have been pressured by our respective lobbies to choose sides. That polarization may obscure the basic fact that we need to be protected from each other in the air. Every airplane is a potential threat to every other airplane. There are, however, some immediate steps that could reduce the threat of mid-air collisions, especially at the busier air terminals.

• Require all aircraft to use landing lights in terminal areas. One study indicated that landing lights were more effective than any available flashers, strobes or beacons for maximizing conspicuity. The Airline Pilots Association (ALPA), a union, and several major airlines have adopted the policy of using landing lights at all times below 10,000 feet.

• Require all aircraft to bear some minimum amount of high visibility paint. Military trainers have been painted in such colors for 40 years. Surely, existing research would indicate that the general aviation fleet should significantly enhance its visibility with proper coloration.

• Require two pilots in all aircraft at high-density airports. Many will take this as a reflection on their own professional abilities, but that would be a mistake. The simple fact is that one pilot sitting on one side of the cockpit just can't see very well out the opposite side. It's a function of aircraft construction as much as pilot ability. If we all had bubble canopies a single pilot might suffice, but until that time I want eyes on both sides of every airplane at O'Hare, Chicago, Newark, or Washington National. If not two pilots, let's at least have a qualified observer to assist that single pilot.

• Require some specific training *and* testing for pilots who use high-density airports. As a minumum, such pilots should spend several hours in a tower and several more in the radar room, observing. Preferably they should be tested on the legal and operational distinctions of ATC procedures in a high density area. Many pilots don't understand the specific differences between instrument approaches, contact approaches, visual approaches, and special VFR.

• Let's get those controllers out of the tower and into some cockpits where they can learn to appreciate the problems of see-and-be-seen. They are welcome in my Boeing 737 anytime, and I hope you will personally invite a controller into your airplane for some familiarization. I know that you both will learn something.

• Utterly abolish all training at major terminals. It is absurd to compound known traffic problems with practice approaches and training maneuvers, which could be more easily and efficiently accomplished at designated training fields. I recall an evaluation flight in an IFR helicopter some years ago during which we executed a practice ILS between DC-10s, at Kennedy International, New York. While I still appreciate the friendly accommodation from Kennedy Approach Control, it does seem unsound that such excursions are allowed.

• Let's have more positive control of aircraft in controlled airspace. Air traffic controllers often are too presumptuous about visual approaches at busy airports, sometimes implying the threat of elongated vectors as the only alternative. Once I was controlled to a position two miles out and 2,000 feet above the Charleston, West Virginia, airport and then instructed to complete a visual approach through marginal visibility. Several other aircraft were milling about in various stages of visual approaches, and the total effect bordered on chaos. Controllers are out of line when they cut corners in marginal visibility. Pilots bear the final responsibility to insist on a complete IFR procedure in such conditions. Our go-around and full ILS back to Charleston, West Virginia, involved 40 nautical miles of vectoring and about 15 minutes of extra flying, but it was the only safe procedure.

Let's introduce some specific training on the very real limits of human vision. The basic concept of see-and-be-seen, which still dominates a large portion of air traffic separation, is predicated on the fallible human eyeball.

Navy pilots used to spend two hours each year on night-vision training, which emphasized the specific problems of seeing at night. Similar training on the normal limitations of human vision in haze, glare and low light levels would help to emphasize the need for vigilance and care.

• Let's accept the responsibility to file reports on all near-misses. Because of the time, paperwork or the seeming threat of legal problems, many pilots are reluctant to follow through with the paperwork after a close encounter. If the NTSB and the FAA heard about every near miss, someone might sit up and listen.

Surely in the next ten years there will be exciting new technical solutions to air traffic conflicts. For the near future, however, let's illuminate our airplanes, learn the system, understand our own limitations and carefully report all of the failures.

SCATANA Is No Joke

16

In the autumn of 1965 a series of mechanical and human failures triggered an electrical blackout over the entire Northeast. Millions of people suddenly were without electricity and several major airports went dark—and quiet—instantly. Pilots approaching those airports at the time tell a chilling story of their fears and problems as they coped with a total failure of the air-traffic-control system. Fortunately, amazingly, there were no accidents.

You may be surprised to know that there *is* a contingency plan for the even more horrendous possibility of war.

SCATANA, or Plan for the Security Control of Air Traffic and Air Navigation Aids, is a joint plan by the Department of Defense, DOT and the Federal Communications Commission. The basic purpose of this three-corner agreement is to provide for appropriate control of civil and military air traffic and navigation facilities in the event of a defense emergency or other threat to national security. It's a contingency plan that may very well never be implemented, but pilots ought to have at least a passing acquaintance with SCATANA's most obvious elements.

The North American Air Defense Command (NORAD) is responsible for implementing SCATANA through its regional commanders. Essentially, NORAD has two options: Emergency SCAT Rules or Full SCATANA.

Emergency SCAT Rules. When an emergency situation develops that does not warrant the declaration of an official Air Defense Emergency, the NORAD regional commander will direct the appropriate FAA Air Route Traffic Control Center to apply Emergency SCAT rules. These rules allow for fluid, real-time restrictions to civil and military flights within any areas so designated by the NORAD regional commander. Flights entering, departing or operating within that area will be issued restrictions or modifications to their flight plans through normal ATC communications.

Basically, VFR flights will be instructed to land at the nearest suitable airport and file IFR flight plans. IFR flights will be assisted by ATC in avoiding or withdrawing from the specified areas. In the meantime, ATC facilities, Flight Service Stations and other appropriate facilities will take action to disseminate to air traffic those instructions and restrictions imposed by the Emergency SCAT rules.

Full SCATANA. When an attack upon the continental United States, Alaska, Canada or allied forces in any area is probable or imminent, NORAD will implement Full SCATANA. If you hear that announcement in flight, pay strict attention because things are going to get worse before they get better.

Full SCATANA requires the immediate grounding of all civil aircraft. VFR flights will be directed to land immediately by blind broadcasts from FAA facilities. IFR flights will receive diversion instructions to appropriate airports by ARTCC. Pilot or passenger convenience is not a factor. You must land as soon as practicable. If you are on an IFR flight, you will be told where ATC will assign that diversion field from one of two lists.

Desired diversion airports are those at which the longest usable runway is of such a length as to afford a *normal* landing for your specific airplane when landing performance is adjusted for the field's elevation.

Minimum diversion airports are those at which the longest runway is of such a length as to afford a *safe* landing for your specific airplane when landing performance is adjusted for the field's elevation.

In both cases, airport requirements are calculated on the basis of maximum allowable structural landing weight and airport conditions of no wind and no runway slope. No approach or go-around limitations are taken into account.

Aircraft are grouped according to runway requirements so that, for instance, G-IIs, DC-4s and Lockheed Electras will all be together. Runway lengths listed for the various categories appear to be realistic under the circumstances, with Falcon 20s scheduled for a desired 5,760 feet and a minimum of 4,990. Those categories and runway lengths are listed in DOT publication 7610.4B entitled "Special Military Operations."

In the worst circumstances you could be ordered to land immediately at an airport with minimum runways and perhaps in an over-gross condition. The natural inclination would be to dump fuel down to some acceptable landing weight, but the extreme situations that dictate Full SCATANA might suggest a very conservative attitude of preserving every precious drop of fuel. Once on the ground at a small, strange airport, it is very likely that fuel, even if available, would be embargoed by the military or hoarded by the proprietor.

Naturally, you will have to make all such decisions based on actual circumstances, but if you elect to land over gross for the prevailing conditions, remember two thumb rules:

• Each one percent of additional gross weight will require one percent of additional stopping distance.

• Each one percent of excess airspeed will require two percent of additional stopping distance.

Before the passengers deplane, inform them of the situation and underline the uncertainty. The offical word will undoubtedly change from hour to hour, and all parties concerned should stay loose and close.

Once Full SCATANA has been established, and when directed by the Joint Chiefs of Staff, further movements of all aircraft will be according to priorities established under the Wartime Air Traffic Priority List for Movement of Aircraft. This list contains 11 separate priorities from "aircraft engaged in active defense of the continental United States" to a catch-all priority for "all other flight operations." Basically the first three categories are reserved for actual combat missions. Categories four through seven are for priority transportation of personnel and supplies as well as for essential administrative flights. Eight, nine and ten are reserved for air-crew training, aircraft testing and administrative logistical flights. Eleven covers everything else. Your movements will depend to a large degree on the circumstances of the moment. You could even end up walking home.

SCATANA is no joke. When that voice in your ear starts assigning airports, cooperate with him. He could save your life.

Handling Overwater Navigational Emergencies

17

A few short years ago "the air traffic control system" effectively stopped at the salt water for most general-aviation pilots. Now, long-range electronic navigational systems and extended-range aircraft allow precise and convenient flight even over large stretches of briny deep. If you don't want to swim, you'd better know how to navigate with and without the convenience.

In ten short years, long-range over-water navigation has undergone a transition from sextant and clock to computerized systems such as INS, Doppler, Omega and VLF. In airline cockpits the navigator's table in the left aft corner has quietly disappeared to be replaced by one more gray jumpseat. In general aviation these simple, accurate long-range systems have ushered in a new era of truly global travel. Even the most difficult navigation problems are reduced to easy keyboard entries so that pilots with no previous long-range experience and a minimum of training can operate virtually anywhere in the world.

Ironically, that marvelous automation introduces a new threat: pilots with no background or training in traditional over-water or hostile-terrain navigation techniques may be far more vulnerable in the event of nav system failure. This problem has already produced a classic accident that will be discussed and reviewed for at least several more years. The unhappy story is short and tragic.

A Sabre 40 en route from Keflavik, Iceland, to Frobisher Bay, on Baffin Island, was using an early model VLF nav system for primary navigation. About 30 minutes from their destination the crew lost that crucial VLF guidance when the U.S. Navy-operated stations were switched off for routine calibration. From that point in space, where VLF guidance was lost, to its fatal impact 65 miles short of Frobisher Bay, the aircraft groped through the night sky, lost and off course. It is painful to fault a fellow pilot after a tragedy, but all the evidence indicates that this professional veteran overlooked some valuable alternative means of navigation when the automatic primary system failed. Webster's defines navigation as "the science or art of conducting ships or aircraft from one place to another—by the principles of geometry and astronomy and by reference to devices designed as aids." In the aviation community, navigation has been reduced to a slavish dependence on aids when in fact real navigation ought to be the sum of all available information.

A Korean Airlines fiasco in Russia, apparently the result of a navigation error, is a case in point. One passenger testified that he knew the airplane was far off the desired heading from the sun's position outside his window, and yet the crew apparently overlooked that basic incongruity to follow the headings of a malfunctioning compass.

Business and commercial pilots should never rely on automated navigation systems as their sole reference. Even the term "primary navigation system" must be viewed in the context of first among many. Primary does not, and should not, mean "only" or "exclusive."

Any long-range flight over hostile surfaces should be carefully planned with all possible means of navigation taken into account. Even the crew of a modestly equipped airplane will have several means of navigation at their disposal. At the bottom line there are always a wet compass, chart, pencil and paper, which can provide the necessary tracking information to complete most flights.

One of the really basic milestones for any long-range flight, the point of no return (PNR), is now frequently ignored. PNR is that time in a flight between two points where the fuel required to reach the destination is the same as that required to return to the departure point. PNR is unnecessary for flights that enjoy the luxury of en route alternatives but it is an invaluable emergency planning aid otherwise. Whenever your aircraft carries just enough fuel to reach its destination (plus reserves), a PNR exists and should be determined.

The formula for PNR is

$$P = \frac{E \times GSR}{GSC + GSR}$$

where

$$P = \text{point of no return in minutes}$$
$$E = \text{aircraft endurance in minutes}$$
$$GSR = \text{groundspeed to return home}$$
$$GSC = \text{groundspeed to continue}$$

A dry-tanks PNR would utilize total endurance. A more practical PNR is derived by using the endurance afforded by planned burnout so that fuel reserves are available at either end. For example, an aircraft with a planned burnout endurance of 300 minutes, cruising at 420 KTAS (Knots of True Airspeed) with 70 knots of average headwind outbound, would fit the formula as follows:

$$PNR = \frac{300 \text{ mins.} \times 490 \text{ kts. GS}}{350 \text{ kts. GS} + 490 \text{ kts. GS}} = 175 \text{ mins.}$$

Up to 175 minutes after departure, this crew have the option of returning to their departure station. After 2+55, they must continue on to the planned destination or alternative.

Actually, this formula lacks some accuracy because it involves averaging of the groundspeed and assumes a constant rate of fuel burn. A much more precise PNR can be obtained by simply working backward from flight-manual and flight-plan cruise data. The engine-failure PNR must be calculated using engine-out groundspeeds. In all cases, PNR will provide you with a valuable decision-making tool for inflight exigencies.

During the flight it is important to maintain a navigation log as a record of the wind and the airplane's track and groundspeed. Then, if the primary navigation system(s) fails, it is an easy matter to establish your position and track to begin the manual navigation problem.

Dead reckoning is a proven method that will lead you to your destination, provided the initial data used to solve the DR problem are correct. Without an up-to-date navigation log, however, you may not be able to determine where you are or what the winds are. Until those data are known, nothing can be done.

The form illustrated here is my idea of a useful long-range navigation log. It is nothing more than a convenient form on which to record all of your navigation information. If all available data are recorded at least every 200 nm, you will have a comprehensive history of your track with which to begin the manual computation. You will also have a performance record for each navigation re-

B/CA Long-Range Navigation Log

Aircraft # _____   Date _____

Time						
Planned Position						
#1 NAV Position						
#2 NAV Position						
Other Position ①						
Line of Position ②						
	NAV 1	NAV 2	Other	NAV 1	NAV 2	Other
Track						
Groundspeed						
Heading						
Drift Angle						
Wind						
Desired Track						

① LORAN (L)   VOR/DME (VD)   Surface Radar (SR)   Weather Radar (WR)   Visual (V)
② VOR, NDB, CONSOL, Standard Broadcast

76

ceiver and a constant awareness of your airplane's position and progress.

Above all, never miss the opportunity to verify your position or line of position (LOP) in relation to any available ground reference. Any information recorded here should also be marked on an aeronautical chart with an appropriate abbreviation to indicate the time and type of fix or LOP. The essential point to remember is that navigation is the sum of all available information. Your VLF, Omega or INS will normally be the most accurate source of information, but you must keep it honest by constantly checking its opinion against any other information. On even the longest overwater legs there are several available supplements to the primary INS, Omega, VLF or Doppler navigation system(s). Consider at least the following:

VOR/DME. It sounds elementary, but even casual use of VOR/DME fixes can reduce your dependence on a long-range nav system by 200 nm or more on each end of the trip. The longest overwater leg in the world without an alternative route is Honolulu to San Francisco, about 2,400 nm. Just using the normal VOR/DME stations at either end of that journey would reduce the long-range navigation problem by 400 nm.

In many cases, it is possible to fix your position by reference to VOR/DME stations that are off-track but within reception range. Plan for that ahead of time and never ignore the opportunity.

Surface radar position. You will not often have this option, but it should be used when available. Across the Greenland icecap, for instance, Sea Bass Radio and Sob Story Radio can fix your position and calculate your groundspeed by reference to military defense radar. From New York to Lisbon, Santa Maria Control in the Azores can give radar advisories. Take them whenever available and record the data and time.

Talking to the correct control facility often results in a false sense of security that can lead you to assume that you are on course. That unfortunate Korean airliner wandering over the Kola Peninsula was apparently lulled by normal communications with an en route controller over 1,000 miles away. Skip and sky wave can do freaky things to communication radio broadcasts, so never assume you are in radar contact merely because you have established voice communications.

Weather radar. The Sabre 40 that was lost southeast of Frobisher Bay probably could have navigated to the field with weather radar. Frobisher is at the end of a long and distinctively shaped bay that shows up very well on airborne weather radar, especially if the ice is out. Airborne weather radar operated in a ground-mapping mode can provide accurate and reliable position-fixing, particularly if the crew is proficient in its use. Coastlines in particular contour well, and this navigation option should be part of any long-range trip procedure. We know of one airline pilot who routinely uses weather radar to track the other airplanes along his Pacific route and employs that admittedly marginal input as one more piece of data to evaluate his INS performance.

Visual position. Along many routes over water or uninhabited territory there are islands, rivers or other prominent surface features that will show positive position-fixing with nothing more sophisticated than the pilot's eyeballs.

When clouds obscure the surface, it is often possible to see the sun, moon or stars, and you don't have to be a celestial navigator to use them for navigation. Polaris, the North Star, is visible all over the Northern Hemisphere and is within one degree of true north. Incidentally, a rank amateur can determine his latitude by shooting the vertical angle of Polaris. It is not precise, but that vertical angle is essentially your latitude. The sun and moon rise in the east and set in the west. It is such basic, commonsense type of information that can eliminate major navigation errors.

Loran. Loran A has been in service since World War II and is a cheap, simple, accurate form of navigation that makes an excellent supplement to more sophisticated systems. Aeroflot, the Russian state airline, routinely navigates the Atlantic with Loran as the primary nav source because the U.S. federal government will not approve its purchase of INS. Eastern Airlines and American Airlines navigate the Caribbean with Loran A.

Consol. Consol is essentially a long-range navigation aid that requires no special equipment. Signals may be received on any low-frequency radio receiver equipped with a beat-frequency oscillator. (Automatic volume control should not be used.)

Consol employs a pattern of alternating dot and dash sectors separated by an equisignal (an area in which two signals of different amplitude merge and become indistinguishable). After the station is tuned and identified, you must wait for one of the silent periods and then count the dots and dashes until an equisignal is heard and continue counting until the next silent period. By careful counting and simple arithmetic, you calculate a final dot or dash count and find your LOP in relation to the station's compass rose on a long-range chart.

OVERWATER NAVIGATIONAL EMERGENCIES

Consol must rank as the world's most tedious navigation aid, but it does work, and at ranges that commonly exceed 1,000 nm. Using Consol requires some prior knowledge of procedure, practice, experience and patience. Instructional material on the use of Consol is scarce, although we were able to find several good descriptions in standard marine navigation texts at the local library. *B/CA* pilots have used Consol over both the Atlantic and Pacific with reasonable results. It's not VOR/DME, but it is better than nothing.

At best, Consol can provide only an LOP, but remember that two intersecting LOPs make a fix. There are at least three other sources of LOP:

VOR. Any VOR radial is an LOP. When DME is available you have a fix, but even without DME that VOR LOP is worth noting and plotting. Remember that variation must be applied at the station's position when plotting VOR LOPs.

Nondirection Radio Beacons (NDBs). Outside of the United States, radio beacons are probably the most common form of navigation aid. The Jeppesen North Atlantic chart is peppered with radio beacons from Canada to Greenland, Iceland and Europe. Any one of these low-frequency facilities can give you an excellent LOP, and two can give you a fix. Remember to apply the variation at your estimated position when plotting an LOP from any radio beacon.

Standard broadcast stations. Standard broadcast station signals are just as useful on the ADF as radio beacons, if you know the stations' locations. Mission Aviation Fellowship has reported tracking radio station KNBR in San Francisco (680 Kilohertz) as far as 600 nm from the California coast. Pan Am pilots routinely listen to the BBC on 200 Kilohertz. That transmitter, about 35 miles from London, reaches out as far as 35 degrees W, or almost to the southwest coast of Greenland. Trans World Radio, on the island of Bonaire, transmits on 800 Kilohertz (and other frequencies) using 500,000 watts of power virtually to blanket the Western Hemisphere with a signal.

When all of the potential sources of navigation information are used, recorded and tracked, you will be able to handle a primary nav failure with ease. The steps are as basic as your first solo cross-country:

- Notify ATC that you may be off track.
- Locate your last known position on the chart.

- Mark your best estimate of present position by plotting the assumed track and groundspeed since that last known fix.
- Make the basic go, no-go decision based on your relation to PNR, the weather forecast, availability of en route and terminal facilities, and fuel. Be sure to consider any available alternatives.
- Plot the desired track to your destination as soon as you have decided where to go. Coordinate any route changes with ATC.
- Calculate the desired compass heading to maintain that track based on variation, deviation and forecast winds.
- Dead-reckon your progress with careful chart notations.
- Try every possible aid to navigation for a fix or at least an LOP. Keep trying until you are sure of your position and in communication with the appropriate ATC facility. You may have as many as several hours of pure dead reckoning, but your only enemies are poor arithmetic, inaccurate wind forecasts or a malfunctioning compass. In normal circumstances, you will be surprised at the accuracy of a simple DR plot.

Long-range navigation over water and uninhabited terrain is now normally accomplished by reference to some form of automated electronic information. Still, the basics of global navigation have not changed since Magellan, and means of establishing your position or LOP are limited only by imagination. Never trust those black boxes any more than necessary; they are generally reliable, but they also are subject to Murphy's Law.

It is an incredible fact that the world has been blanketed with an intricate air traffic control system over land and water, largely during the last three decades. Over water or land it is necessary for every pilot to understand that system in order to avoid the ultimate in-flight difficulty—collision. History underlines the risk:

In 1903 there were two automobiles in the state of Ohio. They collided.

In 1903 there was only one airplane in the world so there were no midair collisions.

Since then the air has been filled with aircraft and an elaborate traffic control system is necessary to minimize the risks of in-flight or ground collision. The system is confusing, frustrating and impersonal but you must understand it in order to fly safely, in order to survive.

ATMOSPHERE, III
THE AIRPLANE'S
HOME

Sunshine is delicious, rain is refreshing, wind braces up, snow is exhilarating; there is no such thing as bad weather, only different kinds of weather.

—John Ruskin

Alas, good old John could afford such a benign attitude toward the weather because he didn't have to fly through all that much. We who earn a living in airplanes naturally adopt a more cynical approach to those problems.

Weather is a constant threat. It is always changing, often contrary, and poorly understood, even by professional meteorologists. Pilots must sort out useful nuggets of information from a landslide of scientific jargon. It takes time, patience, and perseverance.

The United Airlines "Meteorology Home Study Course" contains about 300 closely typed pages of technical content organized into twenty self-taught lessons. It is a valuable course, probably the equal of many college courses. It was mandatory for all new United pilots in the early '60s.

My first four lessons of the MHSC arrived via company mail from our training center in Denver, and I turned to page 1.

"Liquids are nearly incompressible while air is highly compressible." (But does that explain the snow in Muskegon?) "This fact is responsible for many of the behavior patterns we have in nature." (Like how come it's always choppy at my altitude when everyone else is having a

smooth ride?) "Studies of the behavior of gases began a long time ago." (I wonder when they discuss the fog at Salem, Ore.) "We are indebted to Robert Boyle, a British scientist (1659) for the following explanations." (I'll bet old Bob never got rime ice on his horse.) "The following explanation and diagrams should be familiar to you; they are used quite frequently in physics textbooks." (I wonder if they knew how little time I spend perusing physics textbooks?)

Now the truth is that the UAL Meteorology Home Study Course did progress from mind-numbing theory to very useful operational techniques. That was its strength because pilots have a consistent craving for useful knowledge.

Don't tell me about compressibility and dry adiabatic lapse rates, they say. Tell me what those freaky clouds over Pikes Peak mean. Don't tell me about resistance and subsidence. Tell me what that blowing dust on the desert floor near Carson City means. Get the idea? Tell me stuff I can use.

One valuble source of information available to all pilots in all planes is the visual pattern of prevailing weather. In fact, early pilots relied almost entirely on their eyeballs and instincts.

In time those pioneers developed real weather sense although a lot of them were hurt in the process.

Nowadays we rely on satellite pictures, computerized forecasting models and solid-state, digital weather radar. Many pilots still get hurt but the explanations are more lengthy and interesting. Maybe it's time for a renewed interest in those obvious weather patterns which tell their own story.

Sight-Reading the Weather

18

Runways 23L and 23R at Cleveland's Hopkins Airport are within a mile or two of the giant stacks of the Ford Motor Company engine plant in suburban Brookpark. The steam plumes from those stacks provide a continuous visual indication of surface winds during all of the daylight hours. Nevertheless, you can hear one pilot after another asking the tower for a wind check on final approach. Somehow in this controlled and computerized age we have lost touch with nature's own weather indications.

Reports, maps and forecasts have improved steadily during the past twenty years, but weather indications available visually often are the most accurate and current weather data. Every flight should begin with a careful review of all available information, but pilots can supplement that information by sight-reading the weather as it is happening.

Consider steam plumes or smoke trails. Sure, they tell you the wind direction and approximate velocity, but they can do much more than that.

Surface smoke is also a prime indicator of the low-level stability, or resistance to vertical motion, which in turn can be a measure of shear, chop, or fog formation. At least three possibilities exist:

• Smoke or steam that rises steeply into the atmosphere indicates a lack of stability and the possibility of strong convective, or thermal, turbulence—especially during the afternoon hours and thunderstorm season.

• Smoke that flattens out within a thousand feet of the surface is a prime indication of stability, which increases the likelihood of smooth air and light winds on the approach but also creates a general tendency toward haze and fog.

• In all cases, the vertical smoke pattern can indicate levels of changing stability in the atmosphere and wind shear.

• Surface smoke or steam is only one telltale of local weather conditions. Various cloud formations also will offer precise information to the pilot about the chop, turbulence, rain and wind direction that he can expect. They are even useful for predicting the presence of clear-air turbulence (CAT).

CAT is a fact of life at medium and high altitudes. There are, in fact, so many different conditions that produce CAT that forecasters have all but lost their credibility. They issue CAT warnings on such a routine basis that pilots become virtually immune to those stereotype forecasts.

In addition, CAT often threatens such a broad geographical area that even the best forecasts can be frustrating to pilots. Consider the following typical forecast:

Area 1—10,000 to 39,000 feet in area bounded by Allentown, Pennsylvania, Columbia, South Carolina, Memphis, Tennessee, and Fort Wayne, Indiana.

CAT routinely is forecast over such large territories, with a "50-percent probability of moderate or greater." The net result is nearly total apathy on the part of pilots, who cannot effectively use such general information. There has to be a better way, and sight-reading the weather can help.

Much CAT results from wave action that develops at the boundary between two atmospheric layers of differing stability. This sort of wave action can be caused by fronts, jet streams, terrain or the effect of squall lines. The turbulence can cover a few miles or several hundred. The mechanism is similar in each case, and you often can see the signs in distinctive cloud formations.

The wave action begins with a smooth motion called undulance, which may be indicated by a rolling surface on the top of a stratus cloud layer.

As the vertical distance from the troughs to the crests (amplitude) increases, the crest may curl over like the top of an ocean wave because of different wind speeds at that higher altitude. These cresting waves, which often leave a telltale "coat hanger" pattern on stratus layers when they are present, indicate the presence of light to moderate chop.

Occasionally, this kind of wave action will develop into a sort of wave train. Each succeeding crest rises higher than the last until, at some critical height, the largest wave breaks into a chaotic and turbulent area of disturbed air similar to an ocean wave breaking on a beach. Such breaking waves will always produce noticeable turbulence, usually at moderate or greater levels.

Significant wave action is always the result of some contrast in the atmosphere. It is more common in winter than in summer, with the highest frequency in January and February. Summer CAT is most often associated with the blocking action of large thunderstorms or squall lines; it is annoying but usually more local and predictable.

One type of wave action occurs on an enormous scale, with several distinct visual clues. The "mountain wave," as it is called, is a significant producer of turbulence—sometimes violent and destructive turbulence—and you can read the signs in the associated cloud patterns and elsewhere.

When a steep pressure gradient develops directly across a mountain range, the air may be mechanically lifted by the terrain as it naturally flows from high pressure to low. If the lifted air is very stable, i.e., if it has an inherent resistance to vertical changes, the stage is set for mountain wave activity.

The action begins with the mechanical lifting as the predominant easterly flow drives this stable air up the western slopes. At all levels up to 75,000 feet and higher, the air is mechanically held up by the rising terrain. Then, on the lee side, this stable air loses its physical support and cascades down to seek its natural level.

Once the downward motion is begun, it triggers a massive imbalance at all levels, as this basically stable air overshoots and rebounds in several attempts to attain its natural level, like a bouncing ball seeking equilibrium. The size of these vertical motions will greatly exceed the topographical features and the resultant wave action has only recently been measured.

Mountain wave is a common weather feature over the Sierras and Rockies, and at least four visual clues will help you spot it.

Lenticular clouds, named for their unusual lens shape, will form over or near the peaks of the mountains, where the stable air has left its condensation at the peak of its upward journey. Lenticulars are most commonly referred to as "standing"—e.g., "alto cumulus standing lenticular" (ACSL)—although they are very active clouds. Lenticulars are actually in a constant state of formation and dissipation as the western air rises to its peak and leaves condensation before dropping off the eastern slope, drying as it descends. Lenticulars are always active. They will always affect your flight in some dynamic fashion, but they are not always dangerous or even turbulent. You can sight-read that, too.

Smooth, even lenticulars are an indication of smooth, laminar wave action. You will encounter noticeable airspeed variations while flying through the vertical components of this wave action, and you may have to work continuously to maintain a constant altitude, but actual turbulence or chop may not be present at all, at least near the lenticulars themselves. Several layers of lenticulars stacked over the mountain peaks indicate a more pronounced wave action, but if the lenticulars are smooth, they may not be too dangerous. Rough or frayed lenticulars are a sign of turbulent flow and should always be avoided.

But there are other reasons to avoid a mountain wave.

Rotor clouds often will develop on the lee side of the mountains when wave action is present. Such formations are a sign of turbulent or even violent weather.

Rotor clouds are formed as the stable air cascades down the lee side of the mountain, picks up moisture in the lower levels and rebounds to as much as twice the height of the highest peaks.

The clouds formed by this action look like cumulus or fractocumulus lines parallel to the ridge line. Although they appear to be stationary—like the lenticulars—they are actually in a constant boiling motion forming upwind and dissipating downwind. Occasionally, rotor clouds will merge with the lowest lenticulars, which may in turn extend to the tropopause.

Despite their rather benign appearance, rotors are extremely dangerous, with vertical currents often exceeding 5,000 fpm. Positively avoid rotor clouds.

Cap clouds, a form of heavy stratus and fog, will form over the windward slopes and the peaks if there is sufficient moisture in the lower levels. This so-called foehnwal will hang over the lee peaks like a waterfall. The degree of overhang is proportional to the severity of the downdrafts, so you can determine low level turbulence accurately by the extent of the foehnwal overhang.

Blowing dust on a plains surface within 20 miles of the lee slope is a prime indication of surface and mid-level turbulence because the phenomenon marks where the wave has dipped to ground level. A strong wave can lift that dust up to 20,000 feet in an extreme case.

When you encounter any of these visual clues to mountain wave activity, consider your three options:

Detour. Usually the best choice.

Overfly the area several thousand feet above the tropopause, which acts as a barrier to the top of the wave.

SIGHT-READING THE WEATHER

Reduce to rough-air speed, strap everybody and everything down, and hang on.

Wave action, mountain or otherwise, is a prime source of turbulence—so much so, in fact, that meteorologists now prefer the term wave-induced turbulence (WIT) to the older clear-air turbulence. Wave action can produce everything from gentle undulance to chaotic and destructive turbulence, but, fortunately, you often can determine its existence from cloud patterns and surface dust.

Wave action is not the only weather problem you can see, however. There are at least two other common visual clues to turbulence, often encountered in the western United States.

SUMMERTIME AND VIRGA

Virga, a frequent summertime notation on weather reports for Denver and points west, consists of wisps or streaks of water or ice particles falling from the clouds, but evaporating before reaching the ground. The phenomenon usually falls from high-based cumulus and is common on very hot, very dry days. Virga is a lovely weather spectacle but one with fearful implications for the pilot.

Rain falling from these high-based clouds causes extraordinarily strong downdrafts for three reasons.

• The simple mechanical action of the falling water droplets draws air down with them.

• The relatively cool temperature of the falling water chills the surrounding air, causing it to settle even faster.

• The evaporative cooling that results from the introduction of moisture into such a dry environment refrigerates the air even further, accelerating the downdraft.

This triple cooling action accelerates the downdraft into a great cascade, which then spreads out horizontally near the surface. Although these strong vertical currents and gusty low-level outflows will be localized, they can be devastating. Always avoid the area directly beneath virga and be alert for sharp gusts and windshifts when virga is falling near the airport.

There are four conditions you can look for in a weather briefing to help you predict the possibility of virga:

• High-based cumulus clouds.

• Very dry surface air with a temperature-dew point spread of 35°F or more.

87

- Light prevailing winds from the surface to the cloud bases. (Strong winds create a mixing effect that dissipates the downdraft.)
- Temperatures warmer than 75°F.

Any reports of brief gustiness or brief rain showers should serve to confirm the presence of strong downdrafts, when associated with virga.

Another visual phenomenon to look out for is the *dust devil*, a small column of whirling and rising air made visible by the presence of dust and other debris picked up from the dry, hot land surfaces over which it forms. Dust devils usually rise no higher than 1,000 to 2,000 feet, but they are an excellent indication of gross instability at that low level. In fact, dust devils form in a condition of absolute instability, when the higher air is more dense than the lower air and the whole column just swaps ends. The optimum conditions for dust devils are:

- A hot, dry land surface.
- Clear daytime weather with temperatures above 90°F.
- Calm or very light wind.

One visual indicator of incipient conditions is the shimmering effect or "mirage," occasionally observed on very hot, dry days.

Dust devils are benign when compared with their tornado brothers, but they should always be avoided. Any sighting should alert you to the possibility of brief, gusty conditions during the approach or initial climbout phase.

Actually, there are dozens of visual clues to unpleasant weather. Frayed or ragged cirrus clouds are always a sign of rapid mixing in the atmosphere and a prime indicator of turbulence.

Ragged contrails are often a sign of rough air and when there are a lot of these contrails you may be able to pick your best altitude by looking for the smoothest and longest-lasting contrail.

Weather reports and forecasts may give you some early warning of the likelihood of unpleasant weather, but watch for the signs in the sky. Sight-reading these phenomena provides the most current weather report you can get.

Never Trust a Thunderstorm

19

Certainly the single most obvious usual warning of severe weather is the cumulo-nimbus thunderstorm. When you do see one, avoid it. Other methods have been tried but only avoidance will prolong your life.

During World War II, pilots were advised always to fly toward the darkest area of a thunderstorm in order to avoid the worst weather. Later, that advice was amended and flight crews were trained to fly toward the last observed lightning bolt. In the early 1960s, Navy pilots were sometimes requested to fly through thunderstorms at sea for a first-hand report. Now, with the advent of excellent, lightweight airborne radar and improved reporting systems, there is still much confusion about the potential threat of summer thunderstorms.

A single thunderstorm may hold 500,000 tons of condensed water in the form of liquid droplets and ice crystals. When all that vapor condenses into liquid, it releases 300 trillion calories of heat energy. By way of comparison, that much energy is nearly equal to the raw power of a hydrogen bomb. Fortunately, all of that energy in a thunderstorm is not released in a destructive manner, but those big cumulus storms should be treated with great respect.

In the recent past alone, four major air accidents have resulted from flight in or near thunderstorms. A 727 was literally pushed into the ground while on final approach during thunderstorm activity at JFK. Another 727 encountered similar circumstances on takeoff from Denver's Stapleton Airport. A DC-9 at Philadelphia ran out of airspeed in severe wind shear on the edge of a thunderstorm despite the fact that several aircraft had landed safely just prior to that accident. And another DC-9 on descent into Atlanta encountered severe hail that wiped out the compressors of both engines, necessitating a dead-stick landing on a patch of highway.

If there is a common thread to these four accidents, it is this: the ten crewmembers involved willingly and knowingly operated their aircraft in proximity to thunderstorms in the apparent belief that such operations were within reasonable safety limits. It seems obvious that familiarity has bred complacency. Thunderstorms are a serious threat to aerial navigation and must be treated as such. Basically, there are three possible threats associated with any thunderstorm:

Wind shear or turbulence can produce discomfort, passenger injuries, airframe damage and even loss of control. As the thunderstorm builds in the development stage, it creates strong updrafts that fuel the new storm with a steady inflow of moisture. When the storm builds high enough for that moisture to condense and fall as rain, there will be updrafts and downdrafts in close proximity. Near and especially between those opposite vertical drafts will be strong and dangerous shear. Under the cell, the descending air will spread out in all directions to create horizontal drafts or winds with the potential for severe and unpredictable horizontal shear. Any flight through or under a thunderstorm thus will result in passenger discomfort.

Lightning is one of nature's most beautiful phenomena. It also can be very damaging to aircraft structures, fuel tanks and electrical systems.

As the thunderstorm builds, it develops a negative charge center in its base. When this charge center becomes sufficiently intense, it begins to ionize the surrounding air, forming a conducting path. Some of the cloud's negative charge will flow out along this path, forming a charged column of air about three feet in diameter with an intense electrical field at its front. This ionized column grows in a zigzag path called a "stepped leader." As the leader approaches the ground it will attract a similar positive stepped leader, and when the two meet, the conducting path is completed. At that point the return stroke, or lightning flash, occurs, traveling along the leader with a current as high as 200,000 amps and voltages that can reach 100 million.

If your aircraft happens to be somewhere near the path of a stepped leader, it may be electrically attractive enough to be one of the steps. When this happens, the full discharge could pass right through the airplane.

NEVER TRUST A THUNDERSTORM

Hail is possibly the most destructive by-product of a thunderstorm. Those balls of ice, which form at the upper levels of a thunderstorm, are often thrown out of the top or dropped from an extended overhang far from the main cloud. In 1975 the Air Force had five different aircraft that were seriously damaged by hail although none of them was close to a thunderstorm. One F-111 encountered large hailstones while flying in clear air between two buildups 20 miles apart, and a C-130 lost its radome with no significant weather return shown on the radar.

There are other threats from thunderstorms, to be sure, including airframe ice and heavy rain, but shear, lightning and hail should be enough to convince you that thunderstorm avoidance is desirable.

Pilots flying aircraft without radar need to acquire the best possible preflight briefing, including a Digital Radar Code (DRC) plot if available.

Airborne weather radar can provide a level of protection from thunderstorm threats, but it must be used intelligently. Maintain vigilance, using all distance scales in sequence. If you suspect cells in the area and are not painting them use extra gain to force a return. When you do find thunderstorms, avoid them.

Above 23,000 feet, avoid all echoes by at least 20 miles. Below 23,000 feet, avoid steep gradients by at least five miles when the OAT (Outside Air Temperature) is 0°C or higher and by ten miles when the OAT is below 0°C. Increase those distances by 50 percent or more for echoes that are rapidly increasing in size or intensity: changing shape rapidly; or exhibiting hooks, fingers, scalloped edges or other protrusions.

As in all other areas of aviation, judgment and experience should be your guide. Remember that experience indicates that many pilots are taking thunderstorms casually.

Spearheads and Downbursts **20**

One particularly vicious by-product of mature thunderstorms is described by University of Chicago meteorologist Dr. Theodore Fujita as a "downburst." When present, these vertical winds can destroy even large aircraft, and flying beneath the storm may be extra hazardous.

On June 24, 1975, 14 airplanes either attempted to land or landed on Runway 22L at JFK during a 25-minute period. Although the crews flying each of those aircraft reported rain and/or turbulence on the final approach from a nearby thunderstorm, only three were affected in any serious fashion. One pilot reported strong wind shear on close final and forcefully recommended a runway change. Another executed a missed approach because of blinding rain and turbulence. A final one crashed. Surprisingly, other aircraft landed before and in between these three with little or no difficulty.

In a classic case of creative investigation, Dr. Fujita reconstructed the conditions that existed just prior to and at the time of that crash. Using pilot reports, weather reports, radar film from Atlantic City and satellite photographs, Dr. Fujita dissected that thunderstorm into a detailed mesoscale (ten to 100 miles in horizontal dimension) weather analysis. In the process he identified downbursts for the first time. Previously an unknown and violent weapon in the thunderstorm's arsenal, they may be identified by a so-called "spearhead" radar echo.

In the early afternoon of that day, a weak cold front extended from central Pennsylvania to Rhode Island. Due to solar heating, a sea breeze was blowing inland, feeding moisture ahead of the front. At 1400 local time a cumulus line was building in northern New Jersey along the frontal boundary. Over the next 30 minutes the west end of that line grew explosively into a single, towering cumulus. By 1430 the northern end of this line displayed a small bulge. Within 30 more minutes satellite pictures and Pireps (Pilot Reports) revealed a pronounced anvil shape atop this storm as the lifting mechanism spread out in the presence of a relatively stable layer. By 1600, roughly the time of the accident, this cumulus had built into a huge storm complex moving ahead of a small squall line—a sort of rogue storm, always suggestive of extreme weather.

After the accident, there were reports of a tornado-indicating hook echo associated with this storm. Subsequent examination of the radar film revealed no evidence of that hook pattern. There were, however, some unique characteristics to the radar picture.

Initially the storm tracked slowly east-southeast. At 1505 when still about 30 miles from JFK, a spearhead-shaped appendage to the storm began to form on the radar return near the eastern edge of the main echo. This appendage extended eastward very rapidly until it was several miles long. Within half an hour, the spearhead became so large the parent echo started to lose its identity, while at the same time, the storm accelerated abruptly. Finally, the parent was drawn into the spearhead and the storm unified into a single, rapidly moving spearhead echo 15 miles long and five miles wide.

At about 1600 hours that rogue spearhead was positioned just to the north of JFK. Between 1555 and 1605, six airplanes made landing approaches to JFK, including all three that experienced major difficulties. Intrigued by such conflicting experiences in identical airspace in such a short time, Dr. Fujita focused on a space-time analysis of the period surrounding each accident. His analysis indicated that within those ten critical minutes there were three distinct periods of violent weather on the 22L approach path. If the spearhead was traveling at 30 knots, those localized areas of violence within it would be only three to five miles wide. Such extreme localization would account for the capricious incidence of problems along the approach path.

Further investigation revealed that these three critical areas were characterized by extraordinary local outflow. Since such outflow requires a massive supply of descending air, it is clear that they were supported by downdrafts of unprecedented proportions. The concept of downdrafts in thunderstorms is well established; it is defined as "a sustained, non-horizontal current with a downward speed exceeding three fps. (180 fpm)." Since the outflows associated with the JFK storm were of unusual intensity, Dr. Fujita has postulated the new term "downburst" to describe a localized, intense downdraft with vertical currents *exceeding* a downward speed of 12 fps (720 fpm) at 300 feet above the surface. He believes that the three areas of localized violence on the 22L approach path were caused by three individual downburst cells (DBCs) separated by areas of relative calm.

These downburst cells would explain the powerful and chaotic winds reported on and around the airport as the descending air spread out rapidly from the downburst center and the tremendous sink rates and turbulence experienced by the three victimized aircraft on that approach.

Dr. Fujita theorizes that downburst cells originate in the stratosphere. Fed by plentiful moisture and accelerated upward by atmospheric instability the cloud overshoots the stable troposphere above 40,000 feet and penetrates, by momentum alone, several thousand feet into stable, dry, fast-moving stratospheric winds. When the cloud top rapidly collapses, tremendous vertical momentum is imparted to the descending column. This sudden downflow forms a downburst cell on the underside of the cloud and entrains the stratospheric winds that accelerate the downburst away from the main storm.

As the storm's top repeats this cycle of building and collapsing, a succession of downburst cells form and move away from the parent storm. On a radar scope, this family of downbursts might appear as a spearhead echo pointing downward and moving away from the parent. More closely, it might be possible to identify individual downbursts as small, two-to-three-mile circular areas of rain.

There is still much to be learned about downbursts, spearhead echoes, and thunderstorms in general. While there has always been ample reason to avoid one, Dr. Fujita may have discovered one more destructive potential in the thunderstorm riddle.

How the Airlines Use Radar

21

The use of airborne radar by the airlines has passed its twenty-fifth anniversary, a quarter century during which weather radar contributed significantly to air safety and passenger comfort. Today, radar is required by the FAA for all airline operations and is accepted by management and pilots alike as an indispensable aid to flight safety.

During their first 25 years of experience, the airlines developed operating criteria and techniques for the use of weather radar that are applicable to all levels of aviation and all types of radar. Many general aviation aircraft—even singles and helicopters—are now equipped with weather radar. It is a wonderful tool for detecting and avoiding thunderstorms but it is often poorly understood, especially where it is newest.

Effective use of weather radar begins with a clear understanding of what it is. In brief, weather radar is an electronic device for detecting the rate and distribution of rainfall. That's all. Weather radar is nothing more than a rain detector.

Airborne weather radar cannot see turbulence, clouds, wind shear, moisture or lightning. Only water in the form of precipitation will reflect those C- or X-band waves back to the antenna. Even hailstones of significant size are such poor radar reflectors that you cannot expect to paint them, but more on that later. The elemental fact is that your weather radar will only depict rainfall.

But rainfall information is interesting as an *indication* of turbulence and other important weather phenomena, and therein lies the key. It is really guilt by association, or circumstantial evidence.

Meteorologists know that moderate or greater precipitation often is associated with rough air. Therefore, the areas of rainfall painted by your radar are prime suspects to be avoided whenever possible.

Airborne weather radar demands a significant amount of interpretation. Since radar cannot show you turbulence directly, you must deduce the presence of rough air from a knowledge of its correlation with rain. Actually, the equation is primitive: moderate or heavy rain equals rough air. Still, the pilot must be able to infer the presence of that rain from unusual, free-form shapes on his radar scope.

Airline radar techniques are predicated on the use of airborne weather radar as an *avoidance tool*. In general, if the radar shows it, crews are taught to avoid it. There is little or no instruction on how to penetrate radar returns. They all are considered to be potentially harmful.

CORRECT INTERPRETATION IS VITAL

Significant emphasis is placed on storm interpretation as a basis for establishing the avoidance distance. Airline crews are prone to examine a return carefully and then apply some minimum avoidance distance for their particular situation. The four specific conditions that follow are emphasized as warranting particular caution and they often become the airline pilot's chief tools for interpretation.

Steep rainfall gradients, which occur when the water content of a cloud changes rapidly within a short linear distance, mean turbulence.

Really light rain will not even reflect radar pulses back to the antenna, but that is appropriate because light rain normally is not associated with turbulence. Moderate rain will reflect enough signal strength to paint a weak return. As droplet size and number increase the radar image becomes more intense.

When rainfall smoothly varies in intensity from light to heavy over some appreciable horizontal distance, there is little shear effect and probably no serious turbulence. When your radar indicates an abrupt lateral change in rainfall intensity—a steep gradient—from none to intense, you can be sure of significant bumps.

Steep rainfall gradients are indicated by sharply edged radar returns, which depict a strong, localized rainfall. When the radar pictures include bright storm returns with little or no distance between areas of light and heavy rainfall, you are watching a steep gradient. Airline crews tend to avoid steep gradients like the plague, and for a good reason.

Contours are another method of detecting steep rainfall gradients, but their different appearance on radar sets them apart. Early radar had to be interpreted on the basis of image brightness alone. In severe weather it was difficult for pilots to construe adequately the meaning of those bright storm returns so the contour feature was added.

The contour circuit simply excludes all returns of a predetermined intensity. When the contour circuit is in use, really strong returns are specifically blanked out so that contoured areas of storm appear black or hollow. On color radars these areas may appear in a distinctive color, usually red.

Contoured returns are a prime indicator of steep rainfall gradient and heavy turbulence. When the storm return appears as a thin halo of light around a black center, you know that the rainfall gradient is so steep that it builds from nil to contour levels in the brief distance represented by that thin line. Such returns are the worst case, but all contours are to be given a wide margin.

Irregular shapes are a prime indication of fast storm development and the possibility of hail or tornados. Weather radar cannot detect tornados, per se. Neither can your radar paint a useful return from hailstones. Again it's a matter of interpretation.

Ice has less than 20 percent of the reflectivity of water, so hail is normally invisible to airborne radar, with one exception. When hailstones fall through warm air and become coated with water, they become very reflective and will show up as a strong return. Still, there is no easy way to distinguish them from rain. The presence of these two violent weather phenomena, hail and tornados, must be determined from the basic rainfall information available on the weather radar indicator.

Early radar research established a correlation between hooks and scallops on the edge of storm returns and the presence of either hail or tornados. Meteorologists may debate the precise reasons for that correlation, but pilots can take it at face value. Any irregular shaping of radar returns—including hooks, U-shapes or scalloped edges—is a prime indicator of violent weather. Even if they are without any contour, these ragged shapes should be avoided, and by larger than normal distances.

High tops are an important sign of storm development. The antenna tilt control is an excellent tool for interpretation of weather returns, since it has been determined that the greater the vertical development, the higher the probability of turbulence and hail.

These four indicators of violent weather—steep gradients, contours, irregular shapes and high tops—are routinely used to evaluate the potential for severe weather.

KEEP YOUR DISTANCE

Once that evaluation is made, airline dogma calls for avoidance. The following criteria, used by air carriers for detouring around severe weather, should be considered by business and corporate pilots.

• Frequently monitor the longer ranges to assess storm development and to plan early evasive action. Remember that storm returns will intensify as the distance closes because the radar receiver gets a stronger echo from closer targets. Within 50 miles or so, the sensitivity time control circuit equalizes those returns for more accurate interpretation, but early detection is best. When using shorter ranges to detour around echoes, frequently monitor the longer ranges to determine the extent of the area and to watch for additional developments.

When early evasive action is not possible, allow one pilot to fly the airplane while the other concentrates on radar interpretation and detour coordination with ATC. In a particularly active area there is more than enough work for one pilot. If you fly as a single pilot, take some extra precautions even if it means delaying a trip or landing at an alternate.

• In flying between echoes, always choose corridors that are reasonably straight. Crooked, zigzag corridors may be blind traps, created by a radar signal's tendency to be attenuated in the presence of heavy returns. X-band radar is particularly susceptible to attenuation. Look for a straight path through or go around.

• Always avoid flying beneath a cumulus overhang if you have visual contact with the storm. Those anvil tops are prime areas for hail. Many aircraft have been damaged seriously while flying in clear air beneath an overhang.

• Above 23,000 feet, avoid all echoes by 20 nm.

• Below 23,000 feet, airline avoidance policy considers several factors, including static air temperature. Moderate or steep rainfall gradients should be given at least a five-nautical mile berth when OAT is 0°C or warmer and 10 nm when below that temperature. Increase those distances by 50 percent or more for any return that is increasing rapidly in size and intensity, changing shape rapidly or exhibiting hooks, fingers, scalloped edges or other irregularities. Below 23,000 feet, airline policy allows penetration of weak echoes or areas of weak rainfall gradient.

• When you are in the clear on top of clouds, thunderstorms should be overflown by at least 5,000 feet. If it is not possible to clear the tops by that margin, regular avoidance distances are recommended.

• Airline policy discourages the use of gain or intensity controls once they have been set. Some will argue with that stance, but airline crews have experienced little weather damage while flying in accordance with established policy, and that, really, is the bottom line.

• If interference from other radars becomes distracting, try operating in the contour mode. Very strong interference signals may be reduced in intensity by this technique.

Most airlines give crews significant authority to avoid thunderstorms. One major carrier has told its pilots: "If you are in the terminal area with thunderstorms present and you are unable to obtain an ATC clearance in sufficient time to detour a severe thunderstorm which in your judgment threatens the safe completion of the flight, it is appropriate to declare an emergency and advise ATC that you are turning to a heading which will take you clear of the storm."

The airlines have enjoyed extraordinary success in avoiding thunderstorm damage, primarily because their crews faithfully follow firm, stated policies concerning the use of radar. Crew discretion and experience are important, but without the framework of a set policy these qualities can be too indefinite. Corporate flight departments—even single-pilot departments—would do well to specify a set policy for radar use and echo avoidance.

By its very nature, radar is inextricably linked with thunderstorms. That connection engenders two final elements of policy regarding dispatch regulations that are directly applicable to general aviation.

The first forbids any air carrier flight from being dispatched into an area of known thunderstorms with an inoperative radar, except in day VFR conditions. There are no similar restrictions in FAR Part 91 but cancellation should be considered preferable to risking the possibility of encountering severe weather without radar. The second policy involves fuel. Some airlines and pilots routinely fuel for an alternate airport if thunderstorm activity is forecast to be greater than scattered at the destination.

Despite their physical beauty, thunderstorms are incredibly

101

violent and fully capable of destroying your aircraft. A solid radar policy coupled with conscientious radar use, however, can go a long way toward preventing any trouble.

Although it usually is not written into airline policy, weather radar has another excellent use that too often is ignored. In many circumstances, the system can be a solid source of navigational information.

Weather radar is getting better and cheaper. It is available for every class of business aircraft from light singles to heavy jets and many helicopters. Pilots and managers quite naturally will develop a personal feel for this rain detection equipment, but they should consider using the practices, policies and regulations that have served the airlines so well.

Winter Weather

22

Thunderstorms are a spring and summer phenomenon. They are the single most serious threat to aviation safety during those warmer seasons.

Winter problems are very different. One of my friends began his first winter in commercial flying with a serious accident.

We were DC-6 flight engineers then, with responsibility for all pre-flight inspections on that big Douglas transport. Bill was a bachelor with an alert eye for the nicer-looking ladies. That personal weakness and a slippery ramp cost him 14 facial stitches when he skated into a main landing gear door with his attention riveted on a particularly attractive passenger.

Inattention does not always cause accidents but inattention on a slippery surface is particularly hazardous. Bill has the scar to prove it.

Most winter problems are created by some type of freezing precipitation, and extra attention to that information in the sequence reports and terminal forecasts will allow for appropriate advance planning. In effect, you can outpsych winter in the solid comfort of the flight-planning room by developing a clear mental picture of the weather situation along your proposed route.

Remember the very real nature of the hazard of ice and frost on aerodynamic surfaces. Ice will simply change the airfoil to some unknown but less efficient shape. Frost is more subtle. That added roughness on the wing's upper surface increases skin friction and reduces the kinetic energy of the boundary layer. The operational result is an incipient stall, in which flow separation occurs at an angle of attack lower than that of a smooth wing. Frost on the upper wing surface substantially reduces the margins of safety and may, in extreme cases, even prevent takeoff. Don't forget that snow will blow into any available crevice, where it can melt on the relatively warm airframe parts and then refreeze when the temperature drops. Be particularly observant whenever an aircraft has been moved from a hangar into falling snow.

In really cold temperatures it may take awhile to warm the cabin to a comfortable level. If your airplane is pressurized, try this trick to maximize that effort: preheat the cabin by running the heater with all doors closed and the outflow valve open just enough so as not to restrict the warm air flow. When the cabin is warm and when it is time to open a door, close the outflow valve tightly. That action prevents cold air from being drawn in through the outflows as the warm air expands out the door. Airlines use this procedure routinely and are able to maintain comfortable temperatures more than twice as long as they could otherwise.

Allow ample time for your engines to warm up before increasing power above a fast idle. That sounds elementary, but the basic concept has been seriously eroded with the advent of turbine equipment. All engines, turbines included, need to be heat-stabilized to prevent damage and/or reduced output at high-power settings. Regardless of what the flight manual says, allow a minimum of two minutes' running time before using takeoff power. Anything less will detract as much as 15 percent from maximum power and may result in excessive turbine or compressor wear.

If circumstances call for engine anti-ice, turn it on immediately after starting to prevent ice buildup and consequent engine surging during takeoff. In the air, in icing conditions, turn on engine anti-ice systems one at a time with engine igniters on to prevent flameout from the ingestion of ice that has already built up. If you will need anti-ice during the descent, allow extra time and/or distance to compensate for the engine power needed to maintain generator or pneumatic output for the anti-ice circuits.

After prolonged approaches in icing conditions, leave the flaps extended until you can inspect the flap-leading edges for ice buildup that would restrict their retraction, causing a strain on motors, actuators and hinges. Check the landing gear also for any accumulations that might prevent full retraction on departure. And on any departure from slush- or snow-covered runways, cycle the gear at least once to break the ice from doors, actuators and uplocks.

Consider some appropriate adjustment to your operating weights and minimum runway lengths on wet or slippery runways, with and without a crosswind. You already know that runway lengths for takeoff are based on the accelerate/stop distance demonstrated by the manufacturer's best test pilot using a new and flawless airplane on a perfectly smooth and dry runway. Maximum crosswind components and landing lengths are determined in a similar fashion, although FAR Part 25 landing distances are 67 percent longer than the demonstrated landing length.

Water, slush, snow or ice on the runway demands an operational adjustment to the handbook figures. Crosswinds, low visibility, mechanical gripes and minimum experience in type and in season are all grounds for careful personal evaluation of handbook limits. If, for instance, the minimums for your approach are RVR 2400, the maximum demonstrated crosswind for your aircraft type is 25 knots and the minimum runway length for your circumstance is 4,000 feet, it might be best not to try all three at once.

When you do break out on final, the runway may appear only as two black tracks in an ocean of snow. Remember, too, that the far end of the runway is the slickest part of all, especially on the paint marks. The turnoffs are probably even worse, with snow banks to wrinkle the wing tips.

Dig out those engine covers and start using them again, at least when the airplane is left out overnight. They will trap residual heat and allow the engine to cool at a slower and more even rate. Best of all, they keep the machinery from rotating and prevent any foreign objects from accumulating in the nacelles. For example, I once pre-flighted a DC-8 on a cold winter morning and found a neat pile of at least five pounds of sand that had accumulated in each engine intake, including the outboards, which are a good five feet off the ground.

There are lots of tricks for winter flying, but the most effective one of all is simple advance planning. Now go check the snow tires and put a shovel in your trunk. That first sloppy snowfall will be here any day.

Feathered Friends or Fowl Bodies

23

Thunderstorms are the major summer problem, and freezing precipitation is winter's. In between these two seasons pilots are threatened by a different but predictable threat: birds.

It's hard to believe how much damage a single feathery body can do to an airplane in flight until you have seen the results. The windshield of a DC-8 is designed to withstand the impact of a two-pound bird at 400 mph (or a 400-pound bird at two mph). Four hundred mph is about normal climb speed and lots of migrating birds weigh in excess of two pounds so that even the heaviest airplanes are vulnerable to bird strikes.

At FL 180 in level cruise, a Vickers Viscount is struck by one solitary whistling swan. Damage to the central section is immediate and catastrophic. The stricken airliner falls into a vertical dive from which recovery is impossible. No one survives.

A Beech Baron, cruising at 4,000 feet, takes a mallard duck through the wing leading edge. The carcass penetrates far enough to inflict minor damage on the wing spar. Drag from the gaping hole necessitates full opposite aileron at 105 knots indicated airspeed (IAS) to maintain level flight. The aircraft is landed safely by a well-seasoned pilot in good weather on 6,000 feet of runway.

A Lockheed Electra with 67 passengers and a crew of five encounters a flock of starlings 200 feet after takeoff. Number-one engine autofeathers. Numbers two and four experience substantial loss of power. The aircraft yaws to the left, pitches up, decelerates to stall speed and spins into Boston Harbor. Ten people survive.

During the takeoff roll, at very high groundspeed, a DC-10 ingests a number of seagulls into the right-wing engine. As the damaged compressor blades rub against the engine casing, they dislodge and pulverize the interior coating material, which combines with the airflow to form an explosive mixture. The engine disintegrates, damaging the main landing gear and precipitating a disastrous accident that totals the airplane.

A curious seagull lodges in the leading edge slot of a parked 727 when the flaps are inadvertently left extended. On retraction the bird is jammed between the fixed and movable wing parts, requiring an inspection and delaying the trip.

It just doesn't matter where you are. From a parking spot on the ramp, up to FL 200, there is some risk to aviation from these normally benign and enjoyable creatures. There seems to be no way of eliminating the hazards imposed by birds so long as they fly in the same airspace used by those clumsy mechanical imitations we call airplanes. It is possible, however, to minimize the risk with a little knowledge and a few actions.

Simple physics gives some clue to the seriousness of a bird strike. Total force is a function of the bird's weight and impact speed. Double the velocity and you quadruple the force. Double the size of the bird and you increase the force by a factor of five.

If, for example, you strike a one-pound bird at 100 knots, the total impact force would be 880 pounds. If you hit the same bird at 300 knots, the force rises to 7,928 pounds, and at 600 knots, that 16-ounce feathered friend smacks home with almost 16 tons of energy. Similar impacts with a four-pound bird would generate 3,520, 31,712 and 63,200 pounds of force respectively. When you consider that Canadian geese weigh eight to nine pounds, you get an idea of the tremendous potential.

If you must hit a bird, pick a small one and hit him slowly. Especially during climb and descent when exposure is at a maximum and airspeed is largely discretionary, keep it slow. On the way up, that reduced speed will also allow a steeper angle of climb to get you through the worst altitudes quicker. Half of all bird flights are below 500 feet. They fly below 1,500 feet 80 to 90 percent of the time. Above that altitude, birds, like aircraft, are progressively rarer, but bigger.

Those high flyers are normally present only during spring and fall migrations. During those two eight-week periods, hundreds of millions of birds, mostly waterfowl, travel long distances as part of their annual feeding and breeding cycles. Some species prefer daylight operations, navigating by the sun, and others travel at night using star constellations to maintain course. All prefer clear skies for proper navigation, but birds often fly into bad weather unexpectedly so they are not unknown in IFR conditions. Several instances have been reported of iced-up birds falling from over-cast skies. Peak hours of operation are from sunrise until noon and from sunset to midnight although there is likely to be substantial activity at any time.

In the fall, the backside of a cold front provides optimum conditions. The strong northwest winds, cool air and clear skies are perfect for the southerly trip, and birds are well aware of that fact. Ideal conditions occur 24 to 48 hours after a strong Arctic outbreak. Especially when that frontal passage has been preceded by several days of unfavorable weather, there will be a significant increase in migration activity with the front. In the spring, birds must compromise between clear air and favorable winds so that movements are not quite so predictable.

Migration patterns generally follow the ocean coastlines and major inland waterways. Migratory maps show heavy concentrations along the east and west coasts as well as the Mississippi, Missouri, Ohio and other large rivers. Smaller concentrations flow straight south over Salt Lake and along several regional waterway chains in the Central and Eastern states. In fact, the map blankets well over half the country with a much larger percentage of major airports involved since cities tend to be situated near water.

There are several actions you can take to minimize the bird strike potential:

• Take the problem seriously. One bird *can* destroy your airplane.

• Check NOTAMS (Notices to Airmen). During the migration season there are occasional advisories of large sightings.

• Ask for advisories. Ground control may be your best source of information since pilots seem naturally to dump their post-flight comments on that frequency.

• Report presence of birds to others so everyone can share the news.

• Don't be afraid to request a different runway when Jonathan Livingston Seagull is throwing a banquet for his in-laws at the approach end of the active.

• Use windshield heat, landing lights and radar (not to see them, but to warn them) as appropriate.

• Slow down when in the vicinity of reported birds to reduce the impact.

It's true that an ounce of feathers is pretty soft. Unfortunately, when that fluffy ounce is wrapped around a fowl body, it can inflict lethal damage to your aircraft.

Wind Shear and Newtonian Physics

24

One weather phenomenon that is common to all seasons is wind shear, and so is its close relation, wake turbulence. Both of these disturbances have an immediate effect on airspeed, pitch, lift, drag, angle of attack and controllability. There are some rather clear methods for the pilot to forecast or detect the presence of wind shear or wake turbulence. When either or both are present, a single pilot technique applies. That technique is best understood in light of basic Newtonian physics. *Inertia* is the problem. If you can remember that, you will know how to cope with wind shear and turbulence.

It has been over 300 years since Isaac Newton first perceived the basic concepts that led him to his three classic laws of motion. Eight years later Newton's most significant work, the *Principia*, appeared and shook the world of physics. Today, Newton's three laws of motion are recognized as fundamental to the entire science of mechanics.

Over 80 years ago, Otto Lilienthal was killed in a glider crash caused by gusting winds that apparently induced a loss of control. After 2,000 glider flights this brave and inventive pioneer succumbed to a phenomenon that only recently has been labeled "wind shear."

During the past few years, several hundred people have lost their lives in accidents caused by the same problem. And despite a virtual tide of literature on the subject, it still seems to be commonly misunderstood by pilots. But wind shear is largely an inertia problem that Isaac Newton would have understood perfectly.

Newton's first law is familiar to every high-school physics student. "A body at rest remains at rest and a body in motion remains

111

in motion in a straight course unless some force acts on it." This tendency to remain in motion or at rest is called inertia.

To see how inertia works, imagine yourself holding a hammer suspended from a thread. If you jerk the thread quickly, it will break. If you lift steadily and slowly, the hammer will follow without breaking the thread. Inertia is resistance to change, and wind shear is an inertia problem.

Confusion arises from memories of those early ground-school lessons that emphasized that wind does not affect airspeed. That old aphorism is true for steady winds or winds that change gradually, but rapid changes—wind shear—will indeed have a noticeable effect on airspeed, lift, and pitch.

Here, then, is the often misunderstood element: Newton's laws apply to motion *with respect to the earth.* Airplane inertia is, therefore, groundspeed inertia; it is the tendency for an aircraft to maintain its speed over the ground in conformity with Newton's first law. It is, in fact, the physical principle on which inertial navigation systems are based. Rapid wind changes, in conjunction with the airplane's inertia—groundspeed inertia—will cause airspeed fluctuations with possible serious consequences to lift and/or pitch attitude.

Consider an absurd example: an airplane in stabilized, level flight at 80 knots with an 80-knot headwind. Now imagine that the wind instantaneously stops blowing. Because the airplane's groundspeed was zero, its momentum was zero and inertia will try to keep things that way. For the sake of dramatic effect, picture that hovering airplane at 50 feet off the ground just as the wind quits.

Now consider a more realistic example. Picture that same airplane on approach at 130 knots IAS with a 30-knot headwind. Groundspeed is 100 knots, and inertia is acting to sustain that speed differential between the earth and the airplane.

Now imagine that the wind quits entirely. The aircraft's inertia resists any change in groundspeed. With 100 knots of groundspeed and no headwind, the aircraft's airspeed immediately decreases to 100 knots. The pilot is faced with reduced airspeed, loss of lift and probable pitch changes. Aerodynamic forces will tend to reestablish the aircraft in its original, trimmed condition but those forces are relatively slow-acting in relation to that immediate, inertial response to wind shear. Therefore, the pilot must act to correct the situation.

In stable flight, thrust equals drag. Our example aircraft initially had sufficient power to maintain stabilized flight at 130 KIAS. At

100 KIAS, thrust exceeds drag, so the aircraft will start to accelerate; however, that process is relatively slow. Also, because airspeed has diminished, lift is reduced by the square of the velocity change and the aircraft starts to settle. Its nose pitches down, as well, as the aircraft seeks to regain the original trim airspeed of 130 knots.

"Wind in flight does not affect air speed or lift." Remember that from ground school? Yet our imaginary airplane has lost 30 knots or airspeed and nearly 50 percent of its lift, and its nose has pitched down, all due to inertia in the presence of wind shear. Those textbooks should have said, "*Steady* wind in flight does not affect airspeed or lift."

Twenty years ago in Navy primary we were taught to spin a Beech T-34 and then let go, as a demonstration of stability. That stable little trainer eventually would return to its trimmed attitude and airspeed as aerodynamic forces overcame the spinning inertia. It took time and patience, but the maneuver always was begun with enough altitude to ensure recovery. After a wind shear encounter, your airplane will tend to return to its trimmed airspeed and altitude, but during takeoff and approach you may not have enough altitude. As a result, you will have to compensate quickly and effectively for those inertia effects and then be mentally prepared for the next encounter.

A clear understanding of groundspeed inertia is absolutely essential to your understanding of wind shear. Inertia always will tend to keep your airplane at its present groundspeed and direction. Abrupt wind changes will have a significant effect on indicated airspeed as the airplane tries to maintain a constant groundspeed in changing winds.

Wind does not affect airspeed, attitude or lift if the wind is constant, or nearly so. Rapid changes in wind velocity or direction —wind shear—can play havoc with IAS, lift and attitude because of groundspeed inertia.

Now that you understand the problem of inflight inertia, consider your best defense.

Preparation is the essential element in successfully combating wind shear. Although the experts have yet to agree on what are precisely the best piloting procedures for handling extreme shear once it is encountered, they concur that recognizing the nature of the weather problem is the first step in coping with it.

Not all wind shears are caused by exotic downbursts, and not all shear encounters result in *catastrophic* accidents. A shear associated with a seemingly benign front can cause an embarrassing, if not damaging, landing incident; local winds or wave effects can influence an aircraft's speed during approach or departure and increase the risk of a mishap. After encountering shear recently, a Continental Airlines aircraft taking off from Tucson flew through wires located near the departure end of the runway, but the incident received little publicity because the pilot was able to remain airborne, make a successful circuit of the airport and then land safely.

No aircraft is immune to the problems shear can cause, although heavy jets may be more likely than light twins or singles to experience dramatic shear effects due to their higher vertical speeds when traversing existing shears. Still, severe shear effects are not unique to large aircraft.

Wind shear is as much a universal characteristic of weather as other meteorological occurrences, such as low ceilings and visibilities. While shear may not be as easy to detect as zero-zero conditions, and in extreme forms may not exist as frequently, all pilots need to be aware of its character.

Thunderstorms of light to heavy intensity were present in 13 of the 25 accidents that the FAA isolated as possibly involving wind shear in its analysis of NTSB accident and incident reports for the period of 1964 through 1975. Usually, the cells were within a radius of less than 2 nm of the runway threshold and were reported or recorded prior to each accident. Consequently, it is wise to hold off from either an approach or takeoff when thunderstorm activity is in the immediate vicinity of the airport.

One airline recommends holding clear of the intended destination for 15 to 30 minutes to allow thunderstorm-related phenomena to pass by whenever airspeed losses or gains of 25 or more knots have been reported due to shear. It suggests that its pilots delay takeoffs by 30 minutes when wind shears capable of producing 15-knot changes in airspeed are suspected in the area and the conditions are building in intensity. When conditions seem to be dissipating, the same airline recommends a 15-minute delay to allow the thunderstorm's influence to subside.

WIND SHEAR INDICATORS

Even with these rules of thumb, however, the pilot must not proceed in the face of obvious indications that wind shear is still present, such as blowing dust associated with thunderstorm-induced gust fronts or sharp differences in wind velocity and direction at different points on the airport. Studies of thunderstorms have shown that gust fronts—actually, "walls" of high-velocity, low-level wind that reveal themselves by the dust they blow along—can precede the storms that feed them by as much as 30 miles, although a range of five to 15 miles is more typical. That Continental Airlines pilot who encountered the performance-compromising shear on takeoff from Tucson had delayed his departure until a thunderstorm moved off the airport, but his wait was not sufficient to avoid a shear created by the storm's outflow.

Shears caused by fronts are not as easy to anticipate since telltale thunderstorms may be absent, but low-level wind gradients can be present near the frontal zones of stationary, cold or warm fronts. (Occluded fronts generally do not produce shear conditions.) Wind shifts as well as velocity changes close to the surface can occur near the edge of most fronts, but considerable shear should be anticipated when there is a temperature change of 5°C across a front or when the front is moving at a speed of 30 knots or faster.

Other indicators of shear-producing weather are:

Mountain waves. These weather phenomena often create low-level wind shear at airports that lie downwind of the wave. Altocumulus standing lenticular (ACSL) clouds usually depict the presence of mountain waves, and they are clues that shear should be anticipated.

Virga. Precipitation that falls from the bases of high-altitude cumulus clouds but evaporates before reaching the ground is strong indication of low-level wind shear whenever surface temperatures are about 24°C and the dew point spread is 20°C or more.

Strong surface winds. The combination of strong winds and small hills or large buildings that lie upwind of the approach or departure path can produce localized areas of shear. Observing the local terrain and requesting pilot reports of conditions near the runway are the best means for anticipating wind shear from this source.

Sea-breeze fronts. The presence of large bodies of water can create local airflows due to the differences in temperature between the land and water. Changes in wind velocity and direction can occur in relatively short distances in the vicinity of airports situated near large lakes, bays or oceans.

RECOGNIZING SHEAR

The key to recognizing a shear encounter is to know your aircraft's normal performance and power parameters. When the rate of descent on an ILS approach differs from the nominal values for your aircraft, beware of potential wind shear. Since rate of descent on the glideslope is directly related to groundspeed, a high descent rate would indicate a strong tailwind; conversely, a low descent rate denotes a strong headwind. The power needed to hold the glideslope also will be different from typical, no-shear conditions; less power than normal will be needed to maintain the glideslope when a tailwind is present and more power is needed for a strong headwind.

By observing the aircraft's approach parameters—rate of descent and power—the pilot can obtain a feeling for the wind he is encountering. Being aware of the wind-correction angle needed to keep the localizer needle centered provides the pilot with an indication of wind direction. Comparing wind direction and velocity at the initial phases of the approach with the reported surface winds provides an excellent clue to the presence of shear before the phenomenon actually is encountered.

HEADWIND/TAILWIND EFFECTS

The track of the aircraft on the glideslope and the manipulation of power to keep the glideslope needle centered are also specific clues that a shear is being encountered.

Decreasing headwinds. In this situation the wind at altitude is blowing faster or more in line with the runway than is the wind at the surface. As a result the aircraft starts the approach with a lower groundspeed than it has when it crosses over the runway threshold. But remember Newton's First Law of Motion, which states that a body in motion with respect to the ground tries to keep that motion at a constant speed. The aircraft's "ground inertia" resists the aircraft's attempt to increase its groundspeed as the headwind diminishes with altitude. Thus, the aircraft's airspeed falls off about as fast as the wind velocity drops off with altitude, and the subsequent response of the aircraft is a loss of altitude and a nose-down trim change.

In order to correct for this deviation and to recapture the glideslope, the pilot adds power and pitches the nose upward. The increased angle of attack arrests the descent and the added power gradually accelerates the airframe to recover the lost airspeed. The application of pitch and power may result in the aircraft's actually overshooting the glidepath, and a power reduction eventually will be needed to establish the high rate of descent that is required to track the glideslope at the new, higher groundspeed.

This type of wind shear is equivalent to an increasing tailwind; both changes in wind decrease the aircraft's performance. Therefore, a decreasing headwind and an increasing tailwind are often referred to as decreasing performance wind shears.

Increasing headwinds. In this situation the wind vector along the localizer at altitude is of lower magnitude than is the wind vector along the localizer at the runway. (The increasing of the wind's magnitude in a direction of the localizer could have resulted from either the wind's velocity being higher at lower altitudes near the runway or from the wind's being directed more in line with the runway.)

As a result, the aircraft starts the approach with a higher groundspeed than it has when it crosses over the runway threshold. Since the aircraft must lose groundspeed and its "ground inertia" resists that change, the aircraft's airspeed increases practically as fast as the headwind increases. The aircraft responds to this increase in airspeed with an altitude gain and a nose-up trim change.

The pilot corrects for the subsequent ballooning on the glidepath by reducing power and pitching the nose downward. The decreased angle of attack stops the ballooning, and the reduced power gradually decelerates the airframe toward the desired airspeed. But to maintain the glideslope once it has been recaptured, the pilot must add more power than what he was initially carrying because the rate of descent that corresponds with the slower groundspeed will be less than what was required for the initial phase of the approach.

The increasing-headwind situation, which is identical in character to that of a decreasing tailwind, acts to increase the performance of the aircraft. Therefore, this type of wind change is known as an increasing-performance wind shear.

Certainly, these descriptions of decreasing and increasing headwinds are stylized and simplified, but they do illustrate the normal responses of an aircraft to longitudinal wind shear on the approach.

In real life, an aircraft would likely encounter more than one shear level and typically experiences both increasing and decreasing headwinds on the same approach.

By observing the aircraft's airspeed and altitude changes along the glideslope and by being critically aware of the pitch and power manipulations required to track the glideslope, a pilot can detect the existence of shear even when other indicators, such as thunderstorms, are not present.

To understand this point (with the advantages offered by five years of analysis by several agencies and wind data extracted from the aircraft's digital flight-data recorder), consider the approach parameters of the DC-10 that crashed at Boston's Logan International due to wind shear. The aircraft, approaching with its autothrottles engaged, encountered an increasing headwind (actually, the wind went from a strong tailwind to a light headwind), which caused the aircraft to balloon and increase airspeed. The autothrottles reduced power to flight idle in an attempt to recapture the target airspeed.

Then the aircraft entered a slightly decreasing headwind, which caused the airspeed to decrease and a high, apparently undetected sink rate to develop. By the time corrective action was taken the engines could not spool up fast enough to keep the aircraft from striking the approach-light standards and crashing, fortunately with no fatalities.

At no time during the flight was the shear strong enough to compromise the performance of the DC-10—in fact, the aircraft struck the obstructions at an airspeed slightly *above* its reference-approach speed.

Monitoring airspeed and the time required to reach various reference speeds during the takeoff can provide clues to impending shear situations. The time required to accelerate to V_1 will be less than normal in an increasing headwind and greater than normal in a decreasing headwind. Since wind shears (particularly those associated with localized thunderstorms) frequently include both increasing and decreasing headwind situations, unusual variations in the time to accelerate to V_1 and from V_1 to V_2 plus ten knots can be clues to potential shear. Certainly, if there were no headwinds when the takeoff was initiated, but the aircraft seemed to accelerate more rapidly than usual in airspeed between, say, 50 knots and V_1 an increasing headwind is present and shear effects should be anticipated after takeoff.

CONTROVERSY OVER TECHNIQUES

For something as important as establishing the best technique for coping with a strong wind shear encounter, there is a surprising lack of unanimity among knowledgeable parties. Most airline operators and the manufacturers of airline aircraft agree that when conducting an approach in the presence of known decreasing-headwind shear, the pilot should increase his approach reference speed by as much airspeed as he expects to lose due to the shear, but by not more than 20 knots; then, he should attempt to fly a stabilized approach on the normal glidepath. If there are no pilot reports concerning the airspeed loss to be anticipated, the typical formula used by several airlines is to add one half the steady-wind speed and all of the gust speed, with the total not to exceed 20 knots.

For the increasing-headwind situation, where ballooning and an airspeed increase can be expected, pilots of turbine aircraft are cautioned not to reduce power to flight idle in an attempt to increase the rate of descent and to return rapidly to the glidepath. A rough rule of thumb employed by at least one airline is not to reduce thrust to a setting lower than that used on a normal approach when below 400 feet.

Any technique that encompasses approach speeds in excess of V_{REF} runs the risk of an aircraft's crossing the runway threshold with excessive groundspeed and thus running out of runway before it can be stopped. The pilot's option is then to balk the landing and effect a go-around, but the decision must be made early, and the possibility of encountering a shear situation on the initial phases of the go-around, when the aircraft is transitioning from its high-drag landing configuration, must be considered.

Because of the possible need to go around, either because of an overspeed situation at the threshold or due to a performance-compromising shear encounter during the approach, some experts are suggesting that the minimum flap setting consistent with the manufacturer's approved procedures and the available runway length be used where shear is anticipated. However, there is less than complete agreement in this area. A reduced flap setting results in a higher approach speed and a greater exposure to the effects of shear since the aircraft then descends faster through the shear layer.

There also are some differences of opinion concerning what flap setting and climbout airspeed should be used if a wind shear is anticipated during the takeoff. One airline recommends selecting the longest runway and executing a maximum power/minimum flap setting through the rotation, then maintaining a low enough pitch attitude after liftoff to accelerate to V_2 plus 25 knots.

Other parties suggest that the normal takeoff procedure, in

119

which V_2 plus ten knots is used for the initial climb, is better because it provides the advantage of more altitude—which might be needed if a shear caused a loss in climb performance.

The controversy concerning the best speed and configuration for encountering a shear fades into near insignificance when the question of how best to extract an aircraft from an extreme wind shear situation is discussed. The manufacturers and most airline pilots agree that a rapid nose-up pitch correction and an immediate application of power are necessary to arrest a descent that is caused by a severe decreasing-headwind shear and/or a downburst, but there is considerable disagreement concerning how far the airspeed should be allowed to decrease in an attempt to convert speed into altitude.

Manufacturers of airline equipment state that their aircraft possess reasonably good handling qualities and climb performance down to speeds that approach the onset of their stick-shakers, so they are recommending that pilots be aware of stick-shaker speeds and be prepared, in an emergency, to pitch the aircraft's nose upward toward that limit.

Airline pilots, on the other hand, do not want to reduce their speed that far below V_{REF}, preferring instead to leave some airspeed in the bank, so to speak, in case they are forced to flare the aircraft due to impending ground contact.

Other parties, such as the manufacturers of flight-path guidance equipment, feel there are more optimum solutions to tradeoffs between pitch attitude, angle of attack airspeed and longitudinal acceleration that will yield better results when a pilot is attempting to get out of an extreme wind shear situation.

Many manufacturers and pilot-training organizations have no recommended procedure for coping with extreme wind shear.

SO, WHERE ARE WE?

There are some bright spots in the wind shear situation. Here are some things a pilot can do in the cockpit to reduce his risks of a serious encounter with shears and downbursts:

Be prepared. Use all available forecasts and current weather information to anticipate wind shear. The National Weather Service is conducting an experimental program that will lead to inclusion of shear forecasts in regular aviation weather reports, and the Air Force already has such a program operational. Also, make your own observations of thunderstorms, gust fronts and telltale indicators of wind direction and velocity available to other pilots.

120

WIND SHEAR AND NEWTONIAN PHYSICS

Give and request pilot reports. Pireps on wind shear are essential. Request them and report anything you encounter. A suggested format that we feel is worthy of repeating here is as follows:

1. Location of shear encounter.
2. Altitude of shear encounter.
3. Airspeed changes experienced, with a clear statement of
   (a) the number of knots involved, and
   (b) whether it was a *gain* or a *loss* of airspeed.
4. Type of aircraft encountering the shear.

Avoid known areas of extreme shear. When the weather and pilot reports indicate that extreme shear is likely, delay your takeoff or approach.

Know your aircraft. Monitor the aircraft's power and flight parameters to detect the onset of a shear encounter. Know the performance limits of your particular aircraft so that they can be called upon in such an emergency situation.

Act promptly. Do not allow a high sink rate to develop when attempting to recapture a glideslope or to maintain a given airspeed. When it appears that a shear encounter will result in a substantial rate of descent, promptly apply full power and arrest the descent with a nose-up pitch attitude.

Remember that the lift and drag characteristics of an aircraft are such that a change in angle of attack quickly produces more pounds of lift force than it does pounds of drag force over the speed range the pilot will be using during his shear encounter, and those pounds of lift can be used to check the initial impact of a shear or downburst before large descent rates are allowed to build.

Encourage more knowledge. Pilots should seek specific information about the best techniques for extracting themselves from extreme shear encounters. The available information concerning flight below V_{REF} has been generated by the manufacturers of airline equipment for their specific aircraft. Many of their techniques may be type-limited and not applicable to business jets and light aircraft. Considering the simulation and analytical tools available within the business aviation community, it should be possible to develop procedures for the aircraft we fly.

Wake Turbulence 25

Wake turbulence from preceding aircraft—we call it used air—is no more than man-made wind shear. The basic threat is similar, although wake turbulence deserves a special note of its own. Developments in recent years have conspired to underline the serious problems caused by wake turbulence. Airports are more congested, with concomitant pressure on everyone to reduce lateral separations. In addition, more large aircraft are in service.

Fortunately, continuing research programs and heightened pilot awareness have produced new information and helped to establish recovery techniques. But wake turbulence is a permanent problem that bears periodic review.

Wake turbulence is the disturbance generated when an aircraft passes through the air. The total effect includes turbulence from airflow over the fuselage, thrust stream turbulence, lifting surface downwash and wingtip vortices. The strongest and longest lasting elements and, therefore, the most dangerous, are the two vortices trailing the wingtips.

A trailing vortex is a basic result of lift generation by a wing. As you recall, lift is created by the pressure differential of air above and below a moving wing. But the high pressure air below the wing also flows around the wingtip to the low pressure area above the wing, creating a swirling spiral air mass, which trails downstream like a horizontal tornado. Some distance behind the generating airplane the swirls from each wing stabilize into parallel, counter-rotating, cylindrical vortices. Those double corkscrews of high velocity air are remarkable in many respects.

Research has established that trailing vortices settle at a rate of 400 to 500 fpm until they level off at about 900 feet below the flight path of the generating airplane. The turbulence covers a vertical plane about two wingspans wide and one wingspan high. The vortices remain about a wingspan apart, even drifting with the wind, at altitudes greater than a wingspan from the ground. Within about 200 feet of the ground, the vortices spread apart, moving laterally across the surface at about five knots.

A crosswind will decrease the lateral movement of an upwind vortex and increase the movement of a downward vortex. Thus, a light wind of three to seven knots could result in an upwind vortex remaining in the touchdown zone for a period of time or hasten the drift of the downwind vortex toward another runway. Similarly, a tailwind condition can move the vortices of a preceding aircraft forward into the touchdown zone. The light quartering tailwind requires maximum caution.

The intensity of the turbulence is governed by the weight, speed, and wing shape of the generating aircraft. In general, larger aircraft, higher gross weights and slower speeds (higher angles of attack) produce more intense vortices. Wing flaps and leading edge devices produce their own turbulence, which tends to break up the large tip wake. Therefore, the vortices behind an aircraft with flaps extended are more diffuse and tend to dissipate more rapidly than they would if that aircraft were clean. Also, aircraft with T-tails and aft-mounted engines produce greater wake intensities than conventionally configured aircraft of similar size and shape.

Turbulence can reach extraordinary levels inside the wake. Research at the FAA's NAFEC (National Aviation Facilities Experimental Center) facility in New Jersey documented peak vortex tangential velocities of 224 fps, or 133 knots, in the core, with rather uniform reduction in intensity toward the outer limits. Significantly, there was little or no decay of the vortex strength for one full minute and only gradual decay after that. After two minutes, vortex velocity generated by the DC-10 and L-1011 were still about 50 fps, the maximum gust value stipulated by some FARs in determining adequate structural strength.

Safe wake penetration depends on several known elements, which must be mentally juggled for each approach into a wake.

• The size of the preceding aircraft is the most important consideration, although that basic rule has some exceptions. DC-10s and L-1011s generate more intense wake turbulence than C-5s or B-747s, perhaps because of the aft-mounted engines in the first two. Airplanes with T-tails as well as aft engines are comparatively worse.

WAKE TURBULENCE

• Pilots of short-wingspan aircraft must be especially watchful for excessive roll rates. During NASA tests, a Learjet was rolled inverted by the wake turbulence behind a heavy jet, despite full control deflection against the roll. If the wingspan of your aircraft is significantly shorter than that of the preceding aircraft, you may need as much as ten miles' separation to avoid roll rates that exceed your airplane's roll-control capability.

• When following a heavier airplane, the initial wake effect will likely be a sharp rolling tendency. While basic attitude flying demands close attention to wing-level flight, be aware that sharp control movements can induce excessive rolling oscillations as the airplane rebounds between the parallel vortices.

Wake turbulence follows a predictable pattern behind and below the generating airplane, so the rules for avoiding it are simple:

• Always stay above the flight path of the wake generator, whose wake will settle below your path.

• Plan your departure to allow for a liftoff before the point where the heavy lifted off. Use more flap or a modified climb profile, but stay above his flight path.

• Plan your arrival to allow for a landing behind the heavy's touchdown point so that you stay above his flight path.

• Mentally calculate the effect of any crosswind component on the heavy's wake. Depending on the wind's direction and speed, the vortices may hold over the runway or drift toward another runway.

Wake turbulence is heavy stuff. In one accident, a DC-9 was rolled inverted by the wake of a DC-10 and crashed upside down. Imagine what it could do to your airplane.

Airmanship is operating a vehicle in the air. You need to know your vehicle, your ship, thoroughly. You need to understand the air, the medium through which you travel, just as well.

Meteorology is a frustrating and often boring study precisely because it is so nebulous. If you concentrate on the things that do count, on the items that are predictable and that can be compensated for, your aversion to weather-guessing will diminish.

One word of caution. If you learn actually to enjoy studying the weather, keep it to yourself. Real pilots just don't enjoy that. It doesn't fit the image.

PILOTS' RIGHTS, IV PILOTS' RESPONSIBILITIES

It is easy to outline the pilot's responsibility in flight: he is simply responsible for everything. No checklist is required. If anything happens, you are responsible for taking care of it.

It is almost as easy to deal with pilot's rights. If it will enhance the safety of the flight, it is your right. If not, it isn't.

It is classic circular reasoning, when you think about it. The privileges of flight support the responsibilities of flight, which require certain authority. They are inextricably linked.

No one section of a book could properly treat this important subject, but perhaps these few chapters will provoke some thoughts on the crucial subject of professionalism in action.

Passengers 26

Everyone knows about the pilot's responsibilities. That endless list starts accumulating the first day you set foot in an airplane. In fact, it is best to assume that you are responsible for everything and then briefly consider the two or three exceptions that may apply. If you fly airplanes you must be willing to accept responsibility for all the consequences, and some of those lessons can be painful. I remember one in particular.

Ensign Murphy was a bright, likable chap, striving to attain more professional respect and more responsibility with our helicopter squadron aboard the USS *Lake Champlain*. On the day in question, Murphy began his brief career as pilot in command of a big Sikorsky H-34, having finally completed his apprenticeship in the co-pilot's seat. His mission on that fateful day was to retrieve the admiral from the fantail of a destroyer ten miles away. En route to the pickup, Murphy reviewed the procedures for a so-called personnel pick-up.

(1) Hover very low over the fantail so that the hoist operator would need to let down a mere few feet of cable. That would minimize pendulum effects, which could swing the passenger like a trapeze artist.

(2) Hold the helicopter rock-steady so that the horse-collar sling at the end of the cable would hang motionless over the deck as the passenger—an admiral in this case—strapped in.

(3) Lift the passenger gently away and hover over open water momentarily while he is winched into the cabin.

A piece of cake, thought Murphy. I've seen it done a dozen times. A real piece of cake.

The destroyer grew inexorably larger through the windshield as Murphy began the letdown to deck level. He deftly positioned the big helicopter aft on the portside, so as to place the horse-collar sling directly in the middle of the small, cluttered fantail. Everything was going as nice as you please.

True, there was some uneasiness when Murphy realized what an excellent view he was enjoying of the ship's uppermost radio antenna when, in fact, he should have been hovering some 60 feet lower. And the hoist operator did seem to be talking a great deal over the intercom but it was pretty garbled. All that Murphy could make out was, "—slzmch lower melmch swinging bfmky admiral." Also, the whole process was taking entirely too much time, but word finally came over that same intercom that the admiral was in the sling and they were cleared to move off.

Murphy gently eased the helicopter away from the cluttered deck below and initiated a climb for the brief return trip to "Homeplate," the mother ship. At 500 feet and 100 knots the breeze through the hatch precisely compensated for the sunshine through the plexiglass. This pilot was perfectly relaxed and confident when a few simple words shattered his composure, his nerves and his flying career.

"Sir," said the hoist operator over the intercom, "the admiral is not in the cabin yet."

Murphy had forgotten to pause long enough for the admiral to be reeled in and the poor man was now trailing aft on 70 feet of cable like some geriatric Peter Pan. Overriding his own panic, Murphy immediately reduced power and pitched the helicopter to kill speed and retrieve the abused passenger. Maybe he could salvage this mission yet. Everything hinged on a quick enough retrieval of that precious human cargo.

It worked! In fact, it worked so well that Murphy was now able to see, first hand, his second mistake of the day.

Through the windshield, blue sky merged with foam-flecked ocean. Just above that horizon there suddenly appeared a tiny doll-like object, all in white, and seemingly attached to the helicopter by a gossamer thread. Out there on the end of a 70-foot pendulum was the hapless admiral just reaching the apex of his first heart stopping, breath-taking, mind-boggling oscillation.

Murphy had forgotten that his primary responsibility was to that very important passenger. Everything else was subordinate to the admiral's safety and comfort.

It is fair to say that most civil aircraft movements are for the purpose of moving passengers. At the bottom line, after weather and regulations and procedures, your primary responsibility is to protect and deliver those fragile units. It is not always an easy job and the passengers themselves may not cooperate. As pilot in command, you should be familiar with at least the following potential problems:

Drunks. Intoxicated passengers are difficult on any aircraft, but in corporate aviation the problem is often compounded by the very delicate relationship that exists between crew and passengers. Airlines are forbidden by law to accept intoxicated passengers; you may not enjoy that luxury. When you do have a drunk, try to appease him and keep him seated. Consider leaving the seat-belt sign on while setting the cabin altitude at the maximum permissible level and perhaps he will fall asleep. It works quite often.

Upset and angry passengers. This one's a toughie. You cannot allow the passenger to interfere with your judgment or your operation of the aircraft, but his anger definitely needs some attention. The real threat with an angry passenger is that someone will provoke his anger into outright belligerence or irrational behavior. If you can swallow your pride long enough to let him blow off some steam, the problem usually will resolve itself in short order. The primary thing is to avoid a physical confrontation in flight, no matter how much pride you must swallow. If the situation continues to heat up, land.

Problem passengers can complicate a trip even more than many mechanical or operational difficulties. As with most challenges, a small amount of anticipation and planning can turn an otherwise unpleasant experience into a happy—and safe—trip for everyone.

Children and infants. Really small infants are best carried in a bassinet. If possible, position the bassinet against a rear-facing bulkhead during takeoffs, landings and emergencies to protect the child from being tossed about by turbulence or a crash.

The mother will prefer to hold the child in her lap, but the bassinet/bulkhead combination is far safer.

Children often have trouble clearing their ears during climb and descent so it is reasonable to expect that they will cry and complain of a headache or sore ears. Fortunately, crying is one of the best remedies for otitis (inflammation of the ear). When the child begins to cry, he will probably experience some relief. Chewing gum or drinking something may be enough to prevent a serious problem if given at the first sign of discomfort.

If you have children of your own, you know how restless they can get. Simple diversions such as playing cards, pencils, pads of paper, and balloons are usually enough to satisfy most children.

Elderly passengers. This may be the most relative term of all.

PILOTS' RIGHTS, PILOTS' RESPONSIBILITIES

Some people are "old" at 60; my aunt Gert is informed, inquisitive and agile in her 80s.

It is probably fair to say that older people are likely to be somewhat more apprehensive about new experiences than their more youthful counterparts. They will also prefer that you avoid aviation jargon and picayune details in favor of simple, direct explanation.

Blind passengers. There are several definitions of the word "blind," but in all cases, these people lack the vision necessary to function normally. I have carried many blind people over the years and have found them to be interesting, confident and resourceful. All they seem to ask for is a fair shake.

Offer only as much help as the blind passenger desires when boarding and leaving the aircraft. It is too easy to smother him or her with physical maneuvering when simple verbal instructions about steps and doors and seat locations are sufficient.

When he is seated, describe the cabin layout in simple terms including exits, bathrooms, life vests, air vents, ashtrays, reclining seats, refreshment holders, folding tables and call buttons if they are available. Allow the blind passenger to operate the seat-belt and emergency oxygen equipment by himself so he can be independent during the actual flight. When serving a meal, describe the tray layout by the clock method, alerting the passenger to hot items.

Many blind passengers travel with seeing-eye dogs, and this should not present a problem. Don't pat the dog because he will be trained to receive affection only from his owner. Do allow the dog to remain with the passenger, if at all possible. On high-altitude flights arrange the seating so there will be a separate oxygen mask available for the dog.

Deaf passengers. Deafness is a relative term. Experts say there is no such thing as an absolute lack of sensation to all sounds. Even those with so-called total deafness can perceive some extreme frequencies through direct nerve and bone detection. Consequently, deaf passengers will be far more comfortable during the trip if you explain, beforehand, the most obvious changes in vibrations such as takeoff power, landing-gear movements, speed-brake deployment and reverse thrust.

Take time to point out all items of emergency equipment, including the emergency equipment card. It's also nice to explain, with maps, notes or lip reading, the route of flight and other items of interest because your deaf passengers will not hear any inflight announcements.

132

Handling Hazardous Cargo

27

But what if you don't have any passengers? What if you are carrying cargo? And sometimes you will carry both together. I can remember three particular instances of hazardous cargo in my career.

The cockpit of a so-called basic DC-6 included a cargo area that was surrounded by heavy nylon webbing. You could see and hear the cargo in flight because it was not fully closed off.

As a young flight engineer I sat within three feet of this area, from which I could hear a distinct buzzing sound on one particular flight. Closer examination revealed a full load of packaged honey-bees, which were humming away inside their cardboard containers. By the end of that flight the buzzing had driven all three of us, well, buggy. They could have been a real hazard to flight safety if just one of those boxes had been ruptured.

United Airlines carries a lot of airfreight into Denver, and I often co-piloted those trips during my years on the DC-8. Most of the cargo was carried on pallets in the main cabin of those big freighters but the belly pits were also loaded with loose packages.

Those smaller packages were loaded and unloaded by hand so that someone had to crawl into each belly pit and pitch the stuff out onto a conveyor belt. I have a vivid memory of one cargo handler's lightning round trip into the forward pit.

On our way to lunch, we walked right past the conveyor belt as this poor victim opened the pit door and crawled in. He had no sooner disappeared from view than he reappeared in a perfect swan dive as if shot from a circus cannon. His bloodcurdling, airborne scream made it clear that this was no practice dive.

Inside the pit this experienced handler had come face to face with an eight-foot komodo dragon. Someone had not adequately caged this 150-pound lizard, and it was wandering freely around the cargo pit.

Strawberries used to be one of our most common cargo loads from southern California during the season, and they were dangerous for a different reason. First, they smelled so bad in concentration—about like a sick baby—that you could lose your own breakfast by power of suggestion.

Second, once the initial reaction wore off in flight, we ate so many that it was hard to keep our minds on flying. Fortunately, the airline recognized the extreme danger of strawberry overdose and has taken to locking them up tighter than gold bullion. Some say they did that to prevent pilferage but I don't believe it. Safety is always the first consideration.

Still, it grieves me to think that future cargo pilots will not know the joy of sliding over Kansas farmland at 37,000 feet with a big bowl of fresh strawberries in hand.

On a more serious note, many hazardous cargo loads have resulted in real tragedy. One such accident was a milestone.

Sometime in late 1973 a shipping agent in an American chemical plant packed several five-pint glass containers of nitric acid for air shipment. The acid bottles themselves were carefully nestled in sawdust-filled wooden crates with appropriate exterior arrows to indicate "this side up." Subsequently, these boxes were loaded aboard an all-cargo 707 bound from New York to Prestwick, Scotland.

In the main cabin of the Boeing freighter, at least one of those packages was improperly stacked on its side. Sometime during the taxi and climbout, drops of the highly corrosive liquid leaked from the container into the surrounding sawdust, sending up small tendrils of smoke. In very short order, the chemical reaction between nitric acid and sawdust escalated into a destructive and dirty fire, producing enormous quantities of dense, choking smoke.

At first the cockpit crew was aware only of a faint odor. They soon became sufficiently alarmed to request an immediate deviation to Boston. During that increasingly frantic descent, the inaccessible fire raced out of control, producing such opaque and offensive smoke that the captain was forced to open his sliding window for the final approach. Just short of landing, the aircraft was observed to oscillate out of control. It subsequently crashed, killing all three pilots.

The NTSB determined that the probable cause of the accident was the presence of smoke in the cockpit, which was continuously generated and uncontrollable. Although the source of that smoke could not be established conclusively, the Safety Board believes the spontaneous chemical reaction between leaking nitric acid, improperly packaged and stowed, and the improper sawdust packing surrounding the acid's package initiated the accident sequence.

In the airline industry, hazardous cargo has become a hot subject, with no pun intended. In business and commercial aviation, where packaging, handling and loading may be even less professional than the airlines' poor efforts, there is a very real potential for disaster.

If you have occasion to carry cargo in your airplane or helicopter, there are several things you need to know. In fact, there is so much you need to know that we can only hope to pique your curiosity so that you will do some further investigation on your own.

THE RULES

The regulations pertaining to the carriage of hazardous materials on aircraft are contained in FAR Part 103, DOT 49 CFR Parts 170–189 and FAA Handbook 8000.34. FAR 103 is the basic regulation for the transportation of dangerous articles. DOT 49 CFR Parts 170–189 contain supplemental information, including a list of specific dangerous articles exempted from packaging, labeling and inspection requirements. It's available from the U.S. Government Printing Office, Washington, D.C. 20402. The FAA's Handbook 8000.34 describes the authority, responsibilities, policy guidelines and objectives of the FAA in administering the dangerous articles rules. Further, it outlines the procedures Flight Standards personnel use in inspection and enforcement of the provisions of the regulations. It's available from the FAA Office of Information Services (AIS-230), 800 Independence Avenue, Washington, D.C. 20591.

Be fully aware that the regulations pertaining to the transportation of hazardous materials are applicable to *all* civil aircraft (including FAR Part 91 operations) and expressly prohibit the carriage of certain articles in the cabin of a passenger-carrying aircraft. Those hazardous articles that are acceptable for carriage in passenger-carrying aircraft, with the exception of magnetized materials, must be "located in the aircraft in a place that is inaccessible to persons other than the crewmembers."

So, once it's determined that the hazardous cargo can be carried, it must, to be legal, be located in a place where passengers can't get to it. This means that most single-engine business aircraft cannot legally haul hazardous materials at all. In larger business aircraft, it means such cargo must be placed in a baggage compartment that is also not accessible to the crew during flight. Under those circumstances, if a hazardous item starts leaking or breaking, it could go undetected until after a dangerous situation has developed.

Therefore, flight-department managers have good legal reasons for not accepting hazardous material during routine passenger-carrying flights. If requests to carry hazardous materials present a problem in your operations, you may elect to write a prohibition into your operations manual, citing the regulations as your reason.

FAR 103 applies only to medical, beauty and toilet articles in carry-on baggage when those items exceed prescribed amounts. Medical oxygen, small amounts of radioactive material and certain items specified in Title 49 CFR Part 173 are also exempted from restrictions.

But if you do choose to carry items applicable to FAR 103, the rules contain specific labeling, packaging, inspection and quantity requirements. In addition, the FAA must be notified (on a special form) about certain incidents in the course of transportation (including loading, unloading or "temporary storage"), such as the breakage of a container of hazardous material.

The inspection procedure outlined in FAR 103 now requires radiation monitoring of radioactive material before it is accepted for air transportation, as well as monitoring of the compartment in the aircraft *after* the radioactive package has been removed. To perform this inspection, a special monitoring instrument is necessary, along with specialized training in its use. One such instrument is available from Experiments of San Carlos, California. It is a self-contained, battery-operated, portable unit.

HOW TO HANDLE IT

Each potentially hazardous material is unique and requires some very special knowledge and handling by the flight crew. Let's look at some examples.

Compressed gas. Pressurized containers are normally designed for, and filled at, ambient pressures close to that at sea level. When these containers are carried to 7,000 feet, the drop in atmospheric pressure will increase the inside/outside differential by 4.0 psi. At 18,000 feet in a depressurized cabin or baggage compartment, that differential will increase by 7.5 psi. At 39,000 feet it will be 12.0 psi above the sea-level condition. As that differential increases, so does the risk of container failure.

The simplest failure would merely allow the pressurized material to bleed out slowly. Unless the substance in the containers is toxic, corrosive, poisonous or flammable, there would be little or no danger to flight. In an extreme case, the containers could fail explosively, propelling themselves all over the airplane and possibly through any other pressurized containers with which they may have been packed, starting a chain reaction.

To preclude that possibility, it's best to avoid hauling pressurized containers other than cosmetics in luggage. If you must, call the transportation department of the company manufacturing the product and ask about safe ways to transport it by air. In any event, try to isolate the container so that a mishap will not result in a disaster.

Magnetic material. Any large metal object, whether magnetized or not, is capable of distorting your airplane's compass readings. If you carry such items, carefully note the compass indications before and after loading for any discrepancy. When you suspect a problem, take time for at least a cursory compass swing before the flight. Remember, too, that electric motors, generators and audio speakers all contain magnets.

Radioactive material. Radioactive material is routinely used in industry and medicine. For transportation purposes, the radioactivity is measured in Transport Indices (T.I.) at a point one meter from the center of the container. It is required for T.I. to be shown on the label of each individual container. Opinions vary on acceptable maximum T.I. limits, although most airlines prohibit the carrying of any radioactive matter in the cabin during passenger flights. You should establish some specific guidelines for your aircraft, and you may want to start that research with the FAA (Federal Aviation Administration), CAB (Civil Aeronautics Board) and the Atomic Energy Commission. One more note on radioactive material: it is invariably packaged in very heavy containers, which must be well secured in order to protect the aircraft structure and its occupants.

Dry ice. That stuff we call dry ice is really solid carbon dioxide. At sea-level atmospheric pressure this cold solid steadily sublimes to gas. Gaseous carbon dioxide is heavier than air and can be poured downhill just like water. Inside an airplane, and without adequate ventilation, CO_2 will accumulate in puddles from which it may run down to the cockpit during descent and approach. If that happens, the crew could drown when the odorless, colorless gas physically displaces all usable oxygen.

Small amounts of dry ice are an acceptable and useful means of refrigeration. Larger amounts can pose a very real threat. The limitations for your operation must be established in conjunction with the aircraft manufacturer, the critical consideration being ventilation. If you're carrying a large amount of dry ice, it might be wise for at least one crewmember to be on oxygen throughout the flight.

Corrosives. Acids and alkalines in strong concentrations are particularly hazardous. In the event of a leak, they may cause fire, smoke and/or direct corrosive damage to vital aircraft components. They must be packaged, loaded and observed with utmost caution. Even spare batteries should be transported with vigilance and care. Before placing any acid or highly corrosive agent onboard, find out what must be done to neutralize it and be certain you have the required material on board. You don't want to land at some out-of-the-way airport and watch the structure disintegrate for lack of the appropriate neutralizer.

As you can see, this list is incomplete, elementary and fragmented. At a minimum it should include explosives, flammable materials, irritants, poisons and etiological agents. Even such common items as scuba tanks, medical oxygen bottles, fire extinguishers and the boss's dog can turn a routine flight into panic. We know one guy who almost killed himself fighting a bird in the cockpit.

Cargo flights can be a thoughtful adjunct to your transportation services. But before you load that first package, remember this: although passengers may get on your nerves, cargo can get you dead.

Command   **28**

Passengers and cargo may be your most immediate responsibilities on any given trip but over the long term, over a career of flying, every pilot has a primary responsibility to command. It's an age-old challenge.

On August 3, 1492, Christopher Columbus left Palos, Spain, in command of three ships and 90 sailors on an expedition to pioneer a new and shorter route to the Orient. Nine days later that little fleet put in to San Sebastian, in the Canary Islands, for provisions and repairs. On September 6, they sailed westward again, into total isolation for over six months. For almost 200 days those 90 lives were utterly dependent on crude compasses, dead reckoning and the absolute leadership of one determined individual, Columbus himself.

The world has changed since those simpler days of the fifteenth century, and so have the style and form of leadership. Communications have utterly dissolved isolation, which necessitated harsh discipline and control. Navigation has evolved from a mystic art to a routine technology requiring only moderate skill and little thought. And time-to-distance relationships have shrunk from miles per day to miles per minute. Clearly the task of being a captain in the 1980s is vastly different from that of his counterpart in 1492. Only the name and the responsibility remain unchanged.

The name, of course, is unimportant. It is just a title, a symbol, a courtesy. The responsibility is different. It is the cornerstone of command authority and discretion.

> The pilot in command of an aircraft is directly responsible for, and is the final authority as to, the operation of that aircraft.
> ——*FAR Part 91.3 (a)*

PILOTS' RIGHTS, PILOTS' RESPONSIBILITIES

> The concept of command authority and its inviolate nature has become a tenet without exception. The regulations prescribe that the pilot in command, during flight time, is in command of the aircraft and is responsible for the safety of the passengers, crew members, cargo, and airplane. In this regard, he has full control and authority in the operation of the aircraft.
>
> —NTSB

There is no dispute over the absolute authority of the pilot in command. There is, however, some compelling evidence that the authority is not always being exercised with wisdom, or with firmness and judgment. Cockpit command may be the most fruitful area for safety enhancement in the near future of business and commercial aviation.

WHEN AUTHORITY IS LACKING

Consider the trend of professional aviation accidents in recent years as highlighted by the following NTSB accident reports.

• "On September 27, 1973, Texas International Airlines Flight 655, a Convair 600, crashed into the north slope of Black Fork Mountain, Ouachita Mountain Range, Arkansas, while on flight from El Dorado to Texarkana, Arkansas. The crew elected to operate under visual flight rules (VFR) because of frontal activity and associated thunderstorms. The aircraft deviated north of the course between El Dorado and Texarkana and crashed about 80 miles off course.

"Conversations between the captain and the co-pilot, recorded by the cockpit voice recorder, indicated that the crew did not know their position when they initiated a descent from 3,000 feet. About 12 minutes before impact, the co-pilot stated, 'I sure wish I knew where . . . we were.' A few minutes later, he said, 'Painting ridges and everything else, boss, and I'm not familiar with the terrain.' The aircraft descended to about 2,000 feet msl, at the captain's request, while the co-pilot continued to express his doubts about terrain clearance: 'Man, I wish I knew where we were so we'd have some idea of the general terrain around this . . . place.' The captain replied that the highest point in the area was 1,200 feet. Just before impact, the co-pilot had located the aircraft's approximate position, and as he was saying, 'The minimum en route altitude here is forty-four hun—' the aircraft crashed. It struck the mountain 600 feet below the ridge line at an altitude of about 2,000 feet msl.

"The actions of the crew in not using good navigational techniques and their descent when the position of the aircraft was not known must be considered unprofessional conduct."

• "During an instrument approach to Martha's Vineyard, Massachusetts, in June 1971, a Northeast Airlines DC-9 struck the water offshore. A pull-up, initiated just before impact, prevented the plane from crashing. Investigators discovered that the crew did not follow prescribed procedures for altitude monitoring during instrument approaches. None of the required altitude callouts were made because the first officer was busy tuning the low-frequency radio beacon and, on the captain's instructions, he was attempting to contact the company radio for the latest weather reports."

• "In July 1973, a Delta Air Lines DC-9 struck a seawall approximately six feet below the runway elevation at a point about 3,000 feet short of the displaced runway threshold while executing an ILS approach at Logan International Airport, Boston, Massachusetts. The NTSB determined that the accident was partly caused by the crew's failure to monitor altitude and to recognize passage of the aircraft through the approach decision height during an unstabilized precision approach that was conducted in rapidly changing meteorological conditions. During the investigation the Board found that the crew: (1) did not make the required altitude callouts during the approach; and (2) did not abandon the approach on any one of several occasions where deviations were such that a continuation of the approach was unsafe."

• "An Eastern Air Lines Boeing 727-225 struck trees while executing a nonprecision approach to Runway 25 at Toledo Express Airport, in Toledo, Ohio. The Safety Board determined that the probable cause of the incident was the failure of the flight crew to adhere to established procedures when they descended below the authorized minimum descent altitude. The Safety Board concluded from this and other recent accidents and incidents of a similar nature, that inadequate attention to critical operational procedures is a dominant causative factor. It is imperative that the individual pilot recognize the onset of inattention in himself and in others of his crew. It may be combated by adherence to professional standards. These standards must be maintained by alertness, by cockpit discipline, by strict adherence to established procedures."

These and other accidents make it clear that technical competence has far surpassed leadership and judgment in the evolution of airmanship. Forty years ago, Pan American Airways operated

141

S-42 flying boats with 12-man flight crews. Those captains were challenged with frequent equipment failures, inaccurate weather forecasting, elementary navigation equipment and primitive airways and airport facilities. Nevertheless, they were able to exercise rigid authority over 11 cockpit subordinates at a leisurely 85 mph. Their authority stemmed from vastly superior knowledge and experience. It often took the form of rigid autocracy. Respect was absolutely demanded.

TIMES HAVE CHANGED

There have been enormous technical and social changes in the ensuing 40 years. Equipment, weather reporting, navigation and facilities all have advanced to extraordinary levels of dependability. Subordinate crewmembers are normally at a level of technical competence very close to, or equal with, that of the pilot in command. Authority now stems from regulations and is best exercised through example, persuasion, knowledge and reason. Respect is earned. Still, command is just as vital, if not as glamorous.

> Lawful and settled authority is very seldom resisted when it is well employed.
>
> ——*Samuel Johnson*

The legal precept of the pilot in command's authority is the very foundation of flight safety. Other systems have been tried. Very early German flying boats were commanded by a nonflying captain who directed the pilot as if he were merely a helmsman. Some military services have experimented with nonpilot commanders, but it is painfully clear that only the pilot can effectively assume the responsibility for the safety of flight.

Fortunately, the FARs provide all the lawful and settled authority you will ever need to command an airplane, and rightly so. As the pilot, you are at the cutting edge of the action with the clearest perspective for making decisions. Traffic controllers, meteorologists and other crewmembers can be of tremendous help, but you must retain control of the flight. Your challenge will be twofold: to retain the authority that is rightly yours and to employ it well.

> Little is done where many command.
> ——*Dutch proverb*

Authority follows responsibility. That is the equation. Pilots in command are invested with broad authority because they bear

142

sole responsibility for the conduct and safety of flight. The logic is simple, but somehow the issue has become clouded. Authority, as practiced in the cockpit, is under subtle attack from several areas.

In the broadest sense, society itself has not helped. Recent years have seen a headlong rush toward some theoretical state of equality in which no one is more capable or credible than another. Children are taught to question the authority of their parents. Soldiers demand the right to question and criticize orders. Students openly threaten teachers.

Everywhere the individual is encouraged to assert himself beyond all legal or logical limits. In the end, all authority is demeaned.

The aviation community suffers from unique circumstances that may compound this popular denigration of authority. That is, every pilot has tasted command at one time or another, if only in the exercise of simple solo flight. All pilots are quite rightly trained to independence as a means of survival.

Invariably it is the noncaptains who would modify and reduce the essential command authority. Dispatchers often prefer to choose the route, altitude and fuel load. Meteorologists may inconspicuously select the alternate. Air traffic controllers dictate your flight path, sometimes even through squall lines and turbulence. Even other crewmembers readily assume any authority left unattended. It's just a natural human tendency to help, which grows—in a most subtle fashion—to participation and finally to control.

In each case there is a delicate balance between professional support and usurped authority. The seasoned and mature pilot will solicit all available help. He will, at the same time, retain complete authority. That is the key.

The Clipper captain of 1940 had few threats to his authority. As a pilot in command in 1983, you have many such threats, so you need great perseverance in retaining your legal and rightful authority and great skill in exercising it. The style and content of cockpit command is a highly personal and sensitive subject, although some basics endure.

QUALITIES OF COMMAND

Effective command demands four specific qualities essential to the safe operation of an aircraft: knowledge, decisiveness, communication and discipline. Each is as vital to safety as basic piloting skill, and each can be practiced and cultivated by those who aspire to more capable leadership. Leaders may be born, but commanders—cockpit commanders—can be trained and encouraged to a high level of ability.

> If the blind lead the blind, both shall fall into the ditch.
> ——*Matthew* 15:14

Knowledge is certainly the most important prerequisite for capable command in the cockpit area. Knowledge in this context is not mere experience, nor age, nor a collection of certificates and ratings. Command knowledge is the result of a dogged pursuit of useful detail.

Certainly the pilot in command must know his aircraft, environment and regulations. He must also know the applicable weather, alternate options, immediate equipment condition, route and airports for his particular operation. Working knowledge comes from reading, listening, asking and above all recognizing how much we don't know.

Knowledge enhances and solidifies command authority. If you do not read all that is available, if you do not ask, if you are not constantly stocking the memory with pertinent detail and useful theory, you cannot command effectively.

You cannot know everything, even about your own limited area of aviation; your most critical subordinates do not expect that. Still, every correct answer from the individual in authority serves to reinforce his position and each wrong answer or mistake will detract.

You cannot be wrong very often and retain the respect of those who work for you.

> Nothing is more difficult and, therefore, more precious than to be able to decide.
>
> ——*Napoleon*

Decisiveness is utterly critical in commanding an airplane. Consider, though, the distinction between problem-solving and decision-making.

Problem-solving involves the relatively simple process of assembling the available facts and choosing the most appropriate or convenient solution. Problem-solving will lead you to the safest alternate airport, determine the best cruise altitude and resolve most in-flight irregularities. A great deal of this problem-solving is merely choosing the one reasonable course of action that is available. Little if any decisiveness is involved.

144

COMMAND

Decision-making is taking deliberate action when information is so incomplete that the answer does not suggest itself readily. Decisiveness is recognizing alternate courses of action and setting your mind on one. Decisiveness is choosing to go or stay with a marginal magneto drop, and then living with that decision—confident that you have chosen a valid and reasonable course.

As the pilot in command you must make thoughtful but firm decisions and live with them. In fact, the very act of deciding will maximize uncertainty and allow more careful attention to other necessary details of flight. Once the decision is made it must be communicated to all involved in a positive, clear fashion.

> Speak the affirmative; emphasize your choice.
> ——Emerson

Communication is a two-way street: it is the flow of information to and from authority and as such is the hallmark of every organized activity. Precise, intelligent communication is absolutely essential to aviation safety, and yet pilots often play fast and loose with terminology, expression and diction.

As pilot in command it is your responsibility to make your intentions and desires known to ATC, other crewmembers, fuelers, dispatchers, caterers and all who cooperate in the conduct of your flight. You must clearly tell people what you expect of them and you must insist on their clear, verbal response to your directions. Additionally, you must be open to incoming messages so that your vital accumulation of detailed knowledge is not hampered by a closed mind.

Communication is the careful exchange of *knowledge* before and after *decision*. The final ingredient for good cockpit command is discipline.

> *Discipline:* (n.) an orderly or regular pattern of behavior. (v.) to bring under control; to impose order.
> ——*Webster's Third New International Dictionary*

Discipline may be one of the most misunderstood words of the late twentieth century. It is so often misused, in fact, that it may be easier to list the things that it is not. It is not punishment. It is not arbitrary authoritarianism. It is not harsh dictatorship.

Discipline is the firm control and direction of all available resources. It is first personal and only then external. Discipline in the cockpit can be measured by the order, care and completeness of cockpit work. It is most often evidenced by a diligent adherence to standard operating procedures and important details regardless of circumstances.

One of the major threats to discipline is work-load level. Human performance relates to work-load level in a sort of inverted U-shaped curve. At low levels of work load, performance often suffers because of boredom and distraction. High levels of work load may result in poor performance because stress interferes. Optimum performance occurs at intermediate levels of work load, where individuals are challenged but not burdened. Good cockpit discipline, which promotes good order, can even out the flow of work to minimize highs and lows.

Standard operating procedures (SOPs) are important to cockpit discipline. SOPs should include methods of handling the airplane, allocation of crew duties, sequences for checking and operating equipment and checklists for all routine procedures. The SOPs will provide a smooth flow of work load with checks for all critical items and a coherent format that minimizes unnecessary or hazardous actions. SOPs are not a substitute for solid cockpit command, nor are they a hindrance to authority. When properly used, SOPs create the right atmosphere for thoughtful and creative flying.

THOSE NOT IN COMMAND

If it is true, as we believe, that the pilot in command retains absolute authority and responsibility for the conduct of flight, how can those not in command best assist him in that task? Several principles should guide them.

Be confident that your assistance is valuable. As a co-pilot, meteorologist, dispatcher and so on, you are most effective if you have a clear understanding of your own responsibility and if you have developed a positive self-image. You may be playing a supporting role, but your assistance is valuable and in many ways is crucial to safety.

Be ready and able to offer assistance without usurping authority. Your support role can require the tact of a diplomat, the knowledge of a senior captain and the patience of Job. Among the most underrated skills in aviation are those of the co-pilot. He must be flexible, observant, tactful, capable and, above all, helpful. Playing a good support role is challenging and taxing. It may help to know that all great leaders began as subordinates because it takes a good Indian to make a great chief.

Don't confuse respect for authority with respect for a person. Some captains quite naturally inspire personal respect, and they are the easiest ones to assist. Others are cantankerous, egotistical, boorish oafs who inspire little more than revulsion. Most leaders are somewhere between those two extremes, but such personality judgments should not be allowed to cloud the basic concept of authority.

Those in subordinate roles are responsible to the *position* of leadership more so than to the particular *individual* who holds that position. As a co-pilot you may or may not respect your captain, but integrity and professionalism should inspire you to contribute your best job at all times. Never allow a poor leader to degrade your own personal high standards.

Some things transcend mere social change. You, as pilots in command, are the last line of defense in an increasingly complex system of hardware, software, politics, bureaucracy and people. Titles may fall into disuse. (No one remembers Christopher as Captain Columbus.) Styles, methods and personalities change. The job itself will be modified with the advent of new equipment, new techniques and new challenges. Revised regulations may even alter the content of our responsibilities, but through it all I see one pressing need: those in command must preserve that essential but fragile authority for the next generation.

Without it, pilots will become castrated technicians in an impersonal and even dangerous system.

Still, even with extraordinary leadership, things can go awry. And then even if damage is minimal, the reports to your employer, the government and the insurance company will consume a lot of time and energy. While you can't reduce the hurt pride, you can minimize the trauma by preparing in advance for the reports and procedures that must be accomplished after any accident and many incidents.

Since 1967 the sole federal agency responsible for transportation-accident investigation has been the NTSB. The rules pertaining to the notification and reporting of aircraft accidents and incidents are spelled out in Part 830 of the NTSB regulations. This brief, three-page document clearly describes your legal responsibilities, but a more readable and useful pamphlet entitled "Civil Aircraft Accident Investigation Guidelines" is available from the NTSB at Washington, D.C. 20591. This free flyer provides an abbreviated checklist along with excerpts from the applicable regulations.

The NTSB defines an aircraft accident as "an occurrence associated with the operation of an aircraft which takes place between the time any person boards the aircraft with the intention of flight until such time as all such persons have disembarked, in which any person suffers death or serious injury as a result of being in or upon the aircraft or anything attached thereto, or the aircraft receives substantial damage."

A fatal injury includes any injury that results in death within seven days.

A serious injury, according to the NTSB, is "any injury which (1) requires hospitalization for more than 48 hours, commencing within seven days from the date the injury was received; (2) results in a fracture of any bone (except simple fractures of fingers, toes, or nose); (3) involves lacerations that cause severe hemorrhages, nerve, muscle, or tendon damage; (4) involves injury to any internal organ or (5) involves second or third degree burns or any burns affecting more than five percent of the body surface."

Substantial damage "means damage or structural failure that adversely affects the structural strength, performance or flight characteristics of the aircraft, and which would normally require major repair or replacement of the affected component."

There are several specific exceptions to the substantial damage category: engine failure; damage limited to an engine, bent fairings or cowling; dented skin; small, punctured holes in the skin or fabric; ground damage to the propeller or rotor blades and damage to landing gear, wheels, tires, flaps, engine accessories, brakes or wing tips.

Five reportable incidents are specified in NTSB regulation 830.5:
• Flight-control system malfunction or failure.
• Inability of any required flight crewmember to perform his normal flight duties as a result of injury or illness.
• Turbine engine rotor disk failure.
• In-flight fire.
• In-flight collision.

Finally, you must also report any aircraft that is overdue and believed to have been involved in an accident.

Remember, you don't have to be flying or even taxiing in order to have a reportable event, and fatalities and hospitalizations are counted through the seventh day after the fact.

When a reportable event does occur you must make an immediate notification by telephone, telegraph or in person directly to the FAA, who will notify the NTSB for you. That initial notification must include:
• Type, nationality and registration number of the aircraft.
• Name of owner and operator.
• Name of pilot in command.
• Date and time of accident.

- Last point of departure and point of intended landing.
- Position of the aircraft.
- Number of persons on board, number killed and number seriously injured.
- Nature of the accident and weather.
- A description of any explosives, radioactive materials or other dangerous articles carried.

You must make this initial notification by the most expeditious means possible. Supply all the basic information accurately and completely but resist the temptation to speculate about the cause of the accident.

Until the NTSB arrives, you have the responsibility to protect the wreckage along with any mail, cargo or records that were on board. You may not move anything without the express permission of the NTSB except to protect life and prevent further damage. Just let the whole mess lie undisturbed and let the police discourage vandals and souvenir hunters.

You must tolerate reporters, for they have every right to take pictures of the scene and ask questions. On the other hand, they have no right to enter or disturb the wreckage, and you are not obligated to answer their questions.

Any information you are able to record will be very helpful to investigators. If possible, hire a local photographer to document the scene and the wreckage. Note the positions of all pertinent controls and switches, and photograph all the details. Be sure to obtain the names and addresses of any eyewitnesses.

Your final responsibility to the NTSB is to file a written report within ten days. Applicable forms are available from any NTSB office.

Now is the time to establish effective, comprehensive procedures for accident and incident reporting. Those procedures should be coordinated with corporate offices, the company legal department, the insurance underwriter and all flight department personnel.

Accidents are always trouble, but you will be able to react more intelligently if basic preparations have been made.

Short of a definable accident, pilots have the option of reporting problems or unsafe conditions through the Aviation Safety Reporting System (ASRS). These individual and confidential reports are processed and administered by NASA to preclude any possible enforcement action as a result of the report itself.

In fact, the act of reporting an unsafe condition, even if it were a personal violation of the FARs, may provide you with immunity from enforcement should violation be otherwise reported. To avoid

149

a penalty, the offender must be prepared to demonstrate to the FAA that:

1. The violation was unintentional.
2. The violation did not involve a crime or accident.
3. The offender has not been cited and penalized for a violation during the past four years.
4. The offender reported the violation to ASRS within ten days after it happened.

Also, once you receive immunity you cannot claim it for a subsequent violation.

The ASRS is a valuable and viable system for cataloguing and correcting safety problems. It is not a vehicle for reporting accidents but it is the only way you can get a safety message to the FAA without risk of complications, particularly in cases of human error.

Pilot Error 29

Funny thing about the words "human error." They have dogged pilots from the earliest days of aviation, often with an unfair bias. And although I may preach a bit about your responsibilities, I want you to stand up for your rights.

> He was comin' round the ninety,
>   doin' sixty miles an hour.
> When the LSO was heard to scream.
>   Oh they found him in the water
>   With the hand upon the throttle,
> And the mixture in automatic lean.
>
> Now the Pratt and Whitney man said
>   it couldn't be the engine.
> 'Cause the doggone thing would never stop.
>   So what could be fairer
>   Than to call it pilot error,
> 'Cause it couldn't be the doggone prop.

Generations of Navy pilots have sung that cynical ditty at happy hours from Guantanamo to Yokosuku. Unfortunately, the simple convenience of a pilot-error conclusion is often a double tragedy in which the individual aviator's shame masks a very real design or operational problem. Our hero in the water with his hand on the throttle may have experienced a mechanical catastrophe, or an unknown meterological phenomenon, or he may have been guilty of some very human error. It's that human error possibility that is really insidious because it is the most difficult to assess and analyze. Theoretically, there are four types of errors:
- Performance of a required task incorrectly.
- Performance of a proper task at the wrong time.

151

- Performance of an improper action.
- Failure to perform a required task.

Those are the possibilities. For example, you can select the wrong tank. You can select the right tank at the wrong time. You can shut off the fuel altogether or just fail to do anything and run a tank (or all tanks) dry.

But that academic catalogue of errors does not address the more important question of why. I think that there are three classes of pilot error. If we are going to learn anything from the mistakes of others, it is essential to examine their faults in the light of these possibilities.

Design-induced error considers all those design factors that may influence the action, or lack of action, that leads to an accident. One interesting advertisement has claimed that you are statistically safer in a particular jet transport than you would be in your own bathtub. The reason for that is not surprising. The jet transport benefited from a conscious effort to eliminate the potential for design-induced errors, and your bathtub didn't.

Design contributions to human error, even in this era of massive human factoring, is a significant element in many so-called pilot-error accidents. If you will investigate your own operation, you will find several design problems that create the potential for human error. Here are a few I have encountered recently.

Digital counters. Mechanical digital counters suffer from a lack of standardization. DME, altitude, fuel quantity, temperature readings and even time are commonly displayed in this fashion, but with a basic conflict: there is no standard in regard to the direction of movement for increasing values. Some rotate up, some down, and in some cases separate digits in a single readout move in opposite directions.

Levers, knobs, and switches. In spite of standardized efforts in the design of levers, knobs and switches, set-ups for design-induced pilot errors still exist on many flight decks. Look around you at the levers and count the ways inadvertent movements, or movements beyond certain limits, are prevented. Some are prevented by friction (as on certain low-rpm-to-feather tops), some by detent calling for a side motion, others by a thumb latch of one sort or another. Knobs are turned clockwise to increase a value, or decrease it, at the whim of the designer. Identical switches beside one another in a row may be simple on/off, left/off/right, left/both/right or even left/center/right.

Lettering. There is no standard for the form and style of letters, numbers and abbreviations on instruments, panels, placards and checklists. Information is presented in several different typefaces, with random use of upper- and lowercase, numerals mixed with spelled-out numbers and a potpourri of meaningless abbreviations like LT (either left or light) and ALT (either altitude or alternate).

Keyboard design. The trend toward data input through numerical keyboards has accelerated since the introduction of INS nearly ten years ago. Now we have area nav and even total navcom control through these ten-digit keyboards. And there is a conflict.

Many of these avionic keyboards are designed with 1, 2, 3 in the top row; 4, 5, 6 in the middle; and 7, 8, 9 along the bottom. Now look at your digital calculator. Chances are the arrangement is reversed, setting the stage for navigational errors when programming the flight plan on the INS, Omega or RNAV keyboard, and fuel calculation errors when using your calculator. This is obviously unsatisfactory.

Operationally induced error considers those inherent conflicts in the operational environment that contribute to human failure.

• Fatigue is so personal, so difficult to quantify and so important that its contribution to any given accident or incident is almost impossible to assess.

• Terminology has received lots of attention, but it is still a problem.

• Circling approaches demand so much from pilots, controllers and aircraft that the slightest human effort is magnified to disastrous proportions.

• Balanced field-length calculations allow utterly no safety margins for the extreme case so that even predictable human errors may end in tragedy.

And then there is the third category, innate *pilot error.* When it is possible to eliminate design problems, operational factors, and simple acts of God, we are left with the problem of simple human failure. When that is the case, I am not interested in making excuses but I can't help wondering. Was he tired, hungry, sick, distracted, misled, misinformed? What caused him to fail? How many others will repeat that very same error? Most important of all, what can the aviation community learn from that tragic mistake?

Professional Training

30

Effective, professional training is a pilot's inalienable right, if ever there was one. Whether you are employed by a major corporation or fly your own airplane for business purposes you have a professional right to quality training because lives are at stake.

In recent years there has been a virtual revolution in training style and content. The major airlines have been pacesetters in this movement, and their experience is definitely instructive for general aviation because both communities have similar goals. Each of these broad aviation segments must face the classic contradiction of training pilots to a high level of competency but keeping the cost down. The unusual programs developed by major trunk airlines to meet that challenge have had excellent success. It is simply inevitable that the business-aviation community in particular will emulate that experience.

Several trends in airline-pilot training are apparent. It is likely that the shape of business-pilot training over the next several years will be affected significantly by at least three of these trends:

- Greatly expanded use of simulation.
- Programmed learning with tight controls on individual course content.
- Broader curriculum scope.

Pilots often think of simulation only in terms of those cockpit sweat-boxes that have been used as an adjunct to their flight training. In a far broader sense, simulation is a valid educational and research technique that seeks to imitate real phenomena under controlled or artificial conditions. Medical schools, law schools, research institutes and industrial training programs all have made wide use of simulation on various levels. Aviation, especially the space program, has given simulation a broad credibility.

155

Now, after 25 years, the scope and purpose of simulation in flight training are changing with real implications for the serious pilot.

Flight simulation began when Eddie Link built his first "mechanical trainer" in 1929. The Army ordered six units in 1934 and thousands more during the war to teach basic instrument skills. Those old Link Trainers did not attempt to duplicate actual cockpits or flying characteristics, although they were equipped with wings and tails to mollify pilots by giving at least an outward appearance of being airplanes. As instrument trainers they were useful for the introduction of basic procedures. As flight simulators, however, those old Link Trainers were nearly worthless to the pilot.

When the airlines purchased their first pressurized aircraft, they recognized the desirability of more specific simulation. Early Curtiss-Wright Dehmel simulators precisely duplicated the DC-6 cockpit and provided complete system simulation in conjunction with instrument-flight capabilities. Despite their total lack of motion, digital computation or visual presentation, however, those early Dehmel simulators were remarkably valuable adjuncts to flight training; it was 30 years ago that they began the long evolution of flight simulation in professional pilot training.

During the 1960s, flight simulators began to break out of their traditional role as mere accessories to flight training. Cost, safety and convenience combined to force the simulator into a far more demanding role as a total pilot trainer. That trend, which acquired vast credibility from NASA's manned space-flight programs, will continue and will accelerate. Pilots will find that more and more of their training will be accomplished in simulators up to the point of total ATP certification. The complete flight simulator is available now and will become increasingly prevalent.

SIMULATOR ADVANTAGES

Despite the fact that pilots often do not like these torture boxes, simulators offer many advantages in developing and maintaining professional flying skills. For instance:

• Simulators are absolutely safe. Accident records are replete with gruesome training disasters, many of which occurred in spite of a high level of caution and professionalism.

• Simulators are far less expensive to operate than most turbine aircraft, partly due to liability insurance considerations but mostly due to minimal energy requirements.

• The simulator environment, when properly exploited, is notably conducive to the learning situation. First of all, it faithfully imitates the total in-flight experience, including mechanical failures compounded by navigational problems, so that crew coordination is realistically developed. Second, the simulator allows complete attention to the learning process without distractions and inconveniences not helpful to flight training.

• Simulation affords a flexibility and convenience that allow for greatly intensified and more efficient training. Engine failure take-offs, for instance, can be practiced at the rate of about one every three minutes because the simulator can be instantly repositioned electronically.

• Digital simulators allow precise measurement and documentation of a pilot's performance. Grading or evaluation of simulator training sessions and checkrides can be reduced largely to computer-generated reports of the pilot's actual performance. The simulator monitors the pilot's actions and generates a printout of his results. Some major airlines are using this sort of performance measurement as part of their hiring procedure for pilots.

Even visual displays have reached an acceptable level. The newest method of computer-generated imagery employs a computerized inventory of all light locations and characteristics for a given airport as it would appear at dusk or at night. Any number of different airports and/or runways can be called up from the computer memory by the instructor.

Lights can be reproduced in any of seven different colors, and the picture intensity may be varied to simulate distance or low visibility. A horizon glow effect is used to augment attitude reference with either ground lights below or stars above, as appropriate. On short final, the runway surface texture and markings become visible, and even the illuminating effect of aircraft landing lights is duplicated.

I have flown several visual simulators at United Airlines in Denver and Flight Safety International's Falcon Jet installation. The basic picture realism is remarkable, although the picture's response to aircraft maneuvering can be disappointingly artificial.

Visual simulator displays are far beyond the gimmick stage. At the present level of technology visual displays provide an excellent graphic transition from visual flight to instruments, and from instruments to visual. This capability adds a realism that fosters crew coordination. As realistic substitutes for VMC (Minimum Control Speed) aircraft training, current displays lack only some final improvements to the critical computer software. In fact, the

157

newest simulators imitate flight so well that they nearly obviate aircraft flights for transition and proficiency training.

Despite such progress, pilots are prone to feel that simulators do not fly like the aircraft they are designed to imitate. In most cases, this bias was formed from previous experience in poorly designed, equipped and operated simulators. New simulators are a giant step better. In all cases, the pertinent question is the degree of transference.

Training transfer is a classic psychological measurement of the effect of having learned one activity because of an individual's execution of some other activity. In the case of aviation training the key point is whether proficiency gained in a simulator carries over to an airplane, and the answer is clear.

I recently observed a student who held a private certificate and had a total of 600 hours, with only about 50 hours in the past two years and no large-aircraft experience. After 12 hours of instruction in a complete flight simulator, including a full visual presentation, this student's first period in the airplane was remarkable. The first takeoff and landing were completed without instructor assistance, as was all subsequent flying during that first airplane ride. Instrument procedures, aircraft handling, emergencies and all related flying tasks reflected a high degree of competence. Clearly, simulator training prepared this student very well. Psychologists would say that the transference was almost 100 percent.

Airline simulation training is accomplished in three distinct steps. During the early classroom training, animated model boards are used to simulate and illustrate individual system operations. Cockpit controls and indications on the board react like those in the airplane while the illuminated display illustrates each element of that system's functions.

Outside of the classroom, a Cockpit Procedures Trainer (CPT) is used to simulate all aircraft system functioning together but without any flight simulation. Checklists, procedures and engine and system operations are all practiced in this cockpit mock-up as a prelude to flight simulator training.

After several sessions in the CPT the student progresses to the complete flight simulator in which 80 to 90 percent of his "flying" instruction is given. All of this activity with model boards, CPTs and flight simulators contributes in a building-block fashion to the student's total aircraft proficiency. At the end of a one-month transition course the student receives only two hours of actual aircraft flight time prior to the FAA rating check.

Such a curriculum is certainly not as much fun as the old style of flight training, but it is the future, especially for turbine equipment. I have logged over 500 hours in CPTs and flight simulators in 18 years with a major airline. Professional turbine pilots should anticipate similar exposure for the rest of their careers.

Unfortunately, simulation training for the businessman pilot just hasn't happened yet. The main obstacle is price. A full visual simulator sells for at least $2 million so that the cost of operation is far in excess of the cost of operating the real airplane. Without economic justification, there is just not enough incentive to attract the businessman pilot away from his own airplane even though simulation would enhance his level of training.

Still the owner-pilot can use simulation to good advantage. The lower cost, "desk-top" instrument trainers are excellent for practicing all basic IFR procedures. Even the major airlines are evaluating them for use by their second officers in maintaining basic instrument skills. The new portable simulators are so inexpensive and useful that every business pilot should soon have access to one. The fact that these less sophisticated trainers do not faithfully reproduce any particular airplane merely lowers their transference. But at 75 percent transference, for example, each hour of desk-top simulator flight would be as valuable as 45 minutes of real IFR. Certainly, desk-top simulators provide a fair return for the price.

PROGRAMMED LEARNING

Traditionally, classroom training has been delivered by an instructor who may have employed some audio-visual elements to amplify his basic professional delivery. Even the best instructors had their bad days and inevitable personality conflicts, which seriously hampered the transfer of precise, technical information to their students. The trend now is to place the instructor in a role secondary to carefully programmed material.

Much airline training is presently done in small classes of six to eight students. The basic instruction medium is a programmed audiovisual presentation that displays the material in a progressively detailed fashion. At frequent intervals the program interjects review questions, which the student answers with an electronic responder at his desk. The instructor is present only to operate the equipment and to amplify the recorded material if necessary. Even then, his first response may be to read the pertinent section of the canned script before offering any supplemental explanation. Course curricula are absolutely rigid and changes are made at management levels.

Ground-school instruction has, therefore, become nearly as automated as flight simulation, but there are some real benefits for the student, including:

• The advantage of viewing carefully prepared and edited material. Mispronunciations, slips of the tongue and plain mistakes are virtually eliminated. Even that subtle authoritarianism of the recorded sound track adds to the credibility—and thus to the acceptance—of the programmed material.

• A logical, building-block approach that carries through every program and establishes the one, single style and format of instruction to which the student must adjust.

• The lack of personality conflicts to hinder the learning process.

• The avoidance of war stories and personal experiences, which waste precious class time.

• Frequent testing with review questions and automatically graded responses, which provide the instructor with a continuous check of each student's progress.

Naturally there are some drawbacks. Programmed learning is intense and often dry. Programs are edited tightly to contain only pertinent material. If your mind drifts for even a minute, you will have missed some important information. Also, all that prepared information creates a certain dampening effect, which precludes any class enthusiasm and interplay.

At the bottom line, programmed learning presents material in a much more efficient manner. You will learn more in less time, but school will not be much fun anymore. Those days are gone.

ENLARGED CURRICULUM

Airlines generally have recognized that complete pilot training requires far more than basic aircraft transition and proficiency. There should be a long-range career aspect to a pilot's training, and those extra subjects can be as important as basic aircraft knowledge and proficiency. Corporate and owner pilots must usually develop their own sources for that vital extra information on such subjects as radar use and interpretation, meteorology, long-range navigational techniques and procedures, flight physiology and health maintenance, passenger evacuation, ditching and survival, hijack-prevention and procedures, international operations, aerodynamics and flight control, fuel conservation, first aid and pas-

PROFESSIONAL TRAINING

senger care, flight-department management and company policies, FAA regulations, and special- and hazardous-cargo considerations.

Most salaried and businessman pilots will have to construct some sort of personal study program to keep abreast of such extra information. Books and periodicals are your best sources and we include three here to get you started:

Handling the Big Jets, by D.P. Davies. Published in England, and distributed in this country by Pan American Navigation Service, 16934 Saticoy St., Van Nuys, Ca. 91406. This single volume is a gold mine of solid operational information for the serious turbine pilot. Don't let that unfortunate title mislead you. This book should be in every flight department's library (324 pages, $14.95).

Aviation Weather. Published jointly by the FAA and the National Weather Service as Advisory Circular 00-6A. This 1975 edition deals competently with all of the basic weather considerations. It should be owned and read by every serious pilot (SN 050-007-00283-1, $4.55, Superintendent of Documents, U.S. Government Printing Office, Washington, D.C. 20402).

Business and Commercial Aviation Magazine's Pilot Proficiency and Safety Guide. An anthology of articles from previous issues of B/CA, $10. (Write: B/CA Reprints, Dept. L1080, P.O. Box 278, Pratt Station, Brooklyn, N.Y. 11205.)

Business aviation is an expanding and ever more complex profession. The body of information necessary for safe and efficient operation grows each year. Flight department managers and line pilots should take every opportunity to increase their knowledge.

In summary, pilots must cope with a greatly expanding body of knowledge, much more complex aircraft and an increasingly intricate air-traffic-control system. Training and proficiency programs must keep pace with the real needs of professional pilots. Training experts must develop and refine programs. Managers will need to set high standards of personal competence and knowledge. And pilots will have to demand aggressively the information that is vital to their careers.

Checkrides 31

Naturally, when you have completed the training you will face that inevitable checkride. In fact, there are three absolute certainties in a pilot's life: death, taxes and checkrides.

There may be no other professional group in the world that is so thoroughly and repeatedly tested, analyzed and appraised. While we are often required to perform for some critical audience, little thought has been given to the art and science of good testmanship.

The logbook of this checkee contains some 100 entries for checkrides, which, if nothing else, proves an instinct for survival. Building on that experience, one can perceive a distinct pattern that practically guarantees a no-sweat exam. The procedure is effective for written and oral exams and flight checks. En route checks, simulator checks, operations exams, new ratings, added ratings or income tax audits are all easier if you use "the system."

It's tester versus testee and you are a testee. You were destined from your first airplane ride to be tried, tested, quizzed, questioned, scrutinized, probed and challenged. Being a testee has some very specific disadvantages. Checkrides are usually given by pilots who spend considerable time in that pursuit. Checkrides are usually taken by pilots who spend very little of their time in that pursuit.

Examiners often enjoy the challenge of administering a comprehensive test. Pilots invariably dislike the process.

Testers always have a well-defined game plan. Testees are trained, by experience and habit, to assume a submissive role in which they respond mechanically to an examiners' directions.

Get the point? The tester arrives experienced, cheerful and organized. The testee shows up apprehensive, nervous and unwitting.

PILOTS' RIGHTS, PILOTS' RESPONSIBILITIES

First step in "the system," then, is to determine your own strategy; one designed to minimize the tester's advantages and maximize your own resources.

THE GAME PLAN

There is no limit to the possibilities but you must have *some* strategy. Be specific. Make a written list. Mine has five major steps:
- Determine the scope.
- Prepare.
- Rest.
- Cram.
- Take the test.

Scope. It's nearly impossible to prepare for a test until you know with some accuracy what material you are to be tested on. If you are facing an initial Airline Transport Pilot (ATP) certificate ride, the scope will be very broad. You will need to study FARs, weather, flight planning and basic aerodynamics, in addition to aircraft procedures and systems. Add-on ratings will focus more on the specific airplane or task itself.

Company checks will surely include flight department policies and manuals. Route checks normally concentrate on standard procedures and navigation. Whatever the occasion, make every effort to ascertain beforehand the exact nature and content of the check.

When you are able to determine the scope of the proposed exam, you will have scored two important tactical gains:
- You will have maneuvered the tester into an implied agreement, thereby limiting his range of examination.
- You will have narrowed your own responsibility for study and preparation to that official minimum.

Preparation. As you might imagine, this phase is where you win or lose the most ground. There may be hours or months to prepare, but the process is similar in any case.

First thing to do is to gather up the necessary study material so it can be reviewed in some systematic fashion. Make one big pile: flight manual, Jepp manuals, company manuals, study guides. If it looks like a lot, just be thankful that you found out early in the game precisely what you need.

With the material in hand, make a study schedule. Be realistic about your attention span and limit study periods accordingly. Several one-hour sessions salted through a month's time are preferable to two or three whole days. When the schedule is finished,

164

mark your calendar and stick to it. Tell the wife, kids and golfing partner what you are doing and ask them not to disturb you.

One more thing. Pick a place to study that is free from distractions. Try to get away from telephones, radios and friends. I have two personal favorite areas: the public library and an empty airplane. Both are quiet, comfortable and convenient. And both have a certain businesslike atmosphere, easy for study.

Several years ago, Ohio State University undertook a broad program to analyze and treat student academic problems. The result was a learning procedure designed to overcome poor study habits. Our modified procedure has six specific steps: survey, question, read, recite, write, review.

THE SQ2RWR METHOD

Survey. Most manuals and textbooks are organized in a logical fashion with chapter headings and subheadings that clearly outline the content. These headings are the key to your survey. They show the material's organization and point out how the information is assembled and presented.

Use those headings. Thumb right through the chapter or section for additional headings and summaries. Notice any pictures and diagrams, and skim some of the sentences until you have a broad picture of the material involved. When you have learned what the chapter is about, you will know what to expect and have some idea of the key words and concepts.

Question. Immediately after the survey, compose some questions with the idea of finding answers during the reading phase. If the material includes study questions, be sure to review them at this time, even though you will probably not be able to answer them. It may seem futile at first, but questions provide three specific benefits:

• If you keep them in mind, they sustain your interest as you read.

• By looking for answers, you become an active participant in the learning process rather than a passive, submissive pawn.

• The continuous presence of self-testing reinforces your retention of the material as you read it.

When you anticipate questions in this fashion, right from the start, you build a defensive position, which erodes some of the tester's advantages.

First R: Read. Simple, right? Just start at the beginning and read right through. But don't read passively. Challenge the material as

165

if you were buying something. Pause to recall the questions you developed earlier, and don't be satisfied with anything less than complete answers. Read italicized or underlined items several times. They are accentuated for a reason.

Read everything: picture captions, graphs, footnotes, tables and charts. These often contain information that is not mentioned anywhere in the text. Read at a steady pace that is comfortable but consistent, so information is fed to your brain in a steady, coherent stream.

Second R: Recite. Recite? You mean summarize what I have just read out loud? That's exactly what it means. Pause. Recall the main points that have been made and explain them to yourself in clear, concise terms.

If you just read, as you would a newspaper, it is easy to fall into the delusion that you understand and will remember all the material covered. Technical data is just not absorbed that readily. Recitation keeps your attention on the task and allows you to correct mistakes early in the game.

Don't be afraid to allot a substantial percentage of your study time to recitation. One survey indicates that spending up to 80 percent of study time in this fashion is better than straight reading. Most pilots will substantially benefit from 25 to 50 percent recitation time.

Write. Make an outline of the main points you should remember. Use abbreviated words or acronyms that will aid recall. Read your notes out loud, then rewrite the outline using more abbreviations and reducing acronyms to key words. Repeat this step until all the main points have been condensed from paragraph form through sentence form to single words or short phrases.

Third R: Review. Review is the binder that sets what you have learned. Best times for review are immediately after studying and just before the check itself. If you jot down the highlights and pertinent numbers on three-by-five cards as you read, they will be handy and available for these reviews.

Several casual exposures will rivet home the material in an amazing fashion, once you have read, written out, and comprehended the basics.

THE EXTRA MILE

When you've finished the SQ2RWR, keep up the momentum. You may know it all, but you still need to be tried under fire.

Give your kid the flight manual and let him drill you on emergencies while you ride his pogo stick. Have your wife pump you for limitations numbers with the stereo at full volume and a neighbor blowing smoke in your face. Sit in the back seat of your car and draw a sketch of the aircraft's electrical system while your wife drives.

You get the idea. Test your knowledge under simulated stress at least equal to the checkride environment. If you can pull out the answers under fire, you're almost home free, and your confidence factor will climb accordingly.

During the entire preparation period, be sure to take full advantage of every potential for study. Spend a little extra time on routine flight planning and preflighting. Really study each SID, STAR and approach plate as you use them. Pay extra attention to weather analysis. Be sure to find answers for any open questions.

Now there is one more preparation element that needs attention: the Tiger Factor. Checkpilots invariably arrive with a poise and confidence supported by the status of their exalted position. After all, *he* is going to pass judgment on *you*. When you arrive at the appointed hour, you will want all the genuine confidence you can muster. There *are* ways to tweak his self-assurance.

Start doing pushups or jogging a mile each day in preparation. Health and fitness contribute substantially to poise and bearing. Spend some time reviewing your successes and accomplishments. The very fact that you're up for a checkride probably means that someone with authority has confidence in your professional background and ability.

If you are easily intimidated by examiners, start playing a role. Become George C. Scott, Paul Newman, Humphrey Bogart. Get tough while remaining agreeable. Develop a pleasant but aggressive attitude that places you on an equal footing with the tester. After all, you could probably do his job at least as well as he could do yours. You're no patsy and, besides that, checkpilots probably can't do ten pushups.

Rest, Sweet Rest. If possible, schedule your preparation so as to have at least one day of rest before the checkride. Relax. Organize your flight kit. Buy a new tie. Do an extra 50 pushups. Take your wife to dinner. Go to bed early.

The Cram. Cramming has a certain ugly connotation that is really unfortunate. Naturally you cannot make up for wasted time by grinding past midnight on the night before the checkride. However, if your preparation has been progressive and thorough, you can hone an extra-fine edge in one final hour.

On the big day, allow one hour in the morning (after the pushups) for casual review of all the highlights. This is not a study session in any sense, but rather a final brief exposure to the most important areas. Read quickly through the emergency and abnormal procedures. Look over the applicable Jeppesen pages. Breeze through limitation numbers, and look at the notes you have saved on the three-by-five cards. Eat a solid, protein-rich breakfast.

Plan to arrive at the field early. Remember some basic jungle psychology: second guy on the scene is the stranger. If you arrive before the tester, he will be, in a sense, arriving on *your* turf. Score one point for yourself.

When the examiner arrives, greet him with a pleasant firmness that reflects your extensive preparation. Ask him about his style, so you will clearly understand what to expect throughout the check. There is no call to be pushy, but you are a competent professional. He needs to know that you consider yourself more than a passive subject. You will benefit from any clues he gives about timing and content.

THE TEST

During the ride, be positive. You are in command of the airplane, and you should make the decisions until he specifically and directly intervenes to the contrary. One of our all-time favorite characters was an old piston type checking out in his first jet. He had prepared himself to the nines and was ready to *command* the airplane. During taxi-out the FAA inspector asked a question about some item that had already been covered in the pre-taxi checklist. Seizing the opportunity, our hero brought the airplane to a full stop, set the parking brake and ceremoniously pushed his seat back. Then he turned to the examiner and said, "Just one thing. If you're going to ride along this morning, kindly pay attention."

Pretty hard to bust a guy like that.

Remember, too, that the professional checkpilot really does want to get you through. He begins with the presumption that you are capable and competent, or the ride would never have been scheduled. He knows that you are under pressure and has built a substantial allowance for that stress into his appraisal. Plus, he would dearly love to avoid the hassle of letters, memos and official forms that must necessarily accompany an unsatisfactory report.

Fight like crazy the impulse to lapse into a student mentality. Fly the airplane 60 seconds out of every minute. Never relinquish command authority to the examiner until he specifically assumes that role with some verbal indication. Be the captain. Direct the total cockpit workload and be positive about each separate task. Pay attention and understand the examiner's question or requested maneuver. If you don't understand, ask.

When you make mistakes, just keep on going. Never, never point them out to the instructor. If he considers them important enough, he will review them with you at the appropriate time.

Communicate positively with everyone in the cockpit. Issue commands with enough volume to be clearly heard and understood by everybody. If you project some cheer and felicity in your actions, the tester will be forced to recognize that you are at home and comfortable in the air. At the debriefing, hold your tongue. The die is cast and nothing you can say now will modify the examiner's conclusions.

When it's all over, mentally review each item you wish you had done better and savor the highlights. Then go home and put away the pogo stick, but remember where it is for next time.

Personal Health

32

Health maintenance is a very personal but very real responsibility.

Consider, for a minute, your own personal maintenance requirements. Consider them, if you will, with that same critical attitude normally reserved for aircraft hardware.

If you are a perfectly average American, the statistics are way against you. Americans as a whole are 500,000 tons overweight. Just being American increases the probability of heart disease or stroke by 50 percent due to some unknown combination of diet and lifestyle.

Fifty-five percent of *all* deaths involve diseases of the heart and blood vessels. If you're overweight, add 10 percent (to the 55 percent). If you're overweight and over 40, add 30 percent. If you're overweight, over 40 and have a higher-than-acceptable blood pressure, add 45 percent. If you smoke, add a whopping percentage to your category. The over-40 part is irreversible. Fat, smoking, and blood pressure are not.

Which brings up an old airline story:

It seems that a rather corpulent lady waddled to the check-in counter in the festive mood of a typical tourist. This particular portly passenger was going to Fresno to visit old friends. Everything was fine until the airline agent attached a baggage tag to her suitcase, with Fresno's three-letter code—FAT—clearly visible on both sides. He never knew what hit him when her purse caught him on the head, but he has taken to hand-lettering "FRESNO" in those sensitive cases.

Fat is a relative thing, but basically you are overweight if you exceed your doctor's statistical profile for weight for your height. Above that minimum size the chances for premature death by cardiovascular disease, gall bladder trouble, appendicitis and diabetes are disproportionately increased. Fat is removed by eating less and exercising more. It's that simple. With minor adjustments to your habit patterns, you can lose two to four pounds per month. By this time next year you could be back to fighting size.

High blood pressure and other cardiovascular problems are the single most common cause of death in the United States. Those problems should be treated by a physician, but they are *always* aggravated by fat, lack of exercise and smoking. If you follow the prescriptions for overweight you gently reduce this threat. Still, diet and exercise don't treat the smoking problem.

Cigarettes are bad. You can rationalize, agonize and theorize, but the formula is still the same. Butts are bad. At least one person you know, or know about, will slowly choke and gasp to death with lung cancer or emphysema this year. Hundreds of thousands more will succumb to dozens of other diseases caused or aggravated by smoking. If you haven't developed any of those problems yet, don't forget to add 5,000 feet to actual cabin altitudes when calculating oxygen requirements. If you quit smoking, your body will immediately begin to reverse the ill effects and complete the process in about five years. Do yourself a favor. Grit your teeth and quit smoking.

Doctors are unanimous in endorsing the concept of physical exercise on a routine basis. The list of benefits is impressive:

• Fat deposits are reduced and muscle tone improved.

• The respiratory system becomes more efficient, requiring less ventilation for any given workload.

• The oxygen-carrying capacity of the blood is increased, delaying normal fatigue.

• The liver handles sugars more easily and better metabolizes the fatigue-signaling acids produced by exercise.

• Physically fit individuals endure all types of stress better and tend to relax and sleep more readily.

• The heart becomes more efficient, moving more blood with each beat. If you are not in shape, you may be able to reduce your resting pulse by eight to ten beats per minute with regular exercise. That reduction is roughly equivalent to a 300 rpm to 12 percent reduction from cruise power. Ten beats per minute amounts to 14,400 per day and over five million per year.

The question then is, What kind of exercise? The nice answer is that you can choose your own favorite from a very long list. Virtually any exercise that raises your pulse and respiration to a level near double the norm and sustains it for 10 to 15 minutes will produce excellent results. Bowling and golf are great games, but their exercise benefits are near zero because they do not stress the heart and lungs. Do something vigorous. Run, swim, ski, bicycle. Play tennis, handball, squash or whatever you like that makes your heart pound and your body sweat.

Try to pick something that fits your schedule. I have discovered that the only endurance exercise I personally can find time for every day is jogging, in place, indoors. Ten minutes of solid jogging-running, jumping and hopping in place is as good as covering one mile outside. Indoors you can use music, TV, or radio to break the sheer boredom of running.

Perhaps more than anything else, be regular about the exercise. Start out very slowly so as not to get discouraged and increase the total task periodically. Try to draw that happy medium where you are demanding something extra from yourself each time, but not so much that you lose interest. Slowly develop a 20-minute routine that can be done at least five times a week and you will see remarkable results.

As a pilot you have a long list of responsibilities and some rights. Your survival and your career depend on a proper balance of those two.

GETTING THERE

V

Mother always told you to wear your rubbers when it was raining. They did keep your feet dry, but they were a nuisance. Now, 20 years later, the dry feet are worth the small inconvenience.

Dad said not to drive faster than conditions allowed. His advice has acquired remarkable wisdom since those expensive accidents.

The coach harped about proper warm-up prior to the game to prevent cramps and injuries. He seems much smarter now than he did then.

There are many basic rules in the process of getting somewhere in an airplane, and most of them have to be learned the hard way. After all, pilots are a stubborn, feisty lot who relish learning for themselves.

Getting there is largely a process of doing the obvious but doing it faithfully. Wear your rubbers, drive defensively, take enough fuel and don't fly into the ground. It also involves paying attention to the foibles of human communication and to the proper operation of aircraft systems. Part of my plan for flying anywhere is to leave home with a well equipped flight bag. Everything counts toward that final success.

Flight Bags 33

Every airline in the United States is required to have a flight operations manual that details that company's approved procedures for planning and executing a flight. Every airline pilot carries a flight ops manual in his flight bag and completes a written examination on its contents each year.

The manual itself is divided into several sections including general rules, flight preparation, emergency procedures, regulations and en-route procedures. Surprisingly, that en-route section, which details all of the company's procedures for getting from here to there, is one of the smallest. Somehow there is a subtle feeling at the more serious levels of aviation that once the trip has been scheduled and planned, the actual flight will take care of itself. Often that is the case in good weather and well-maintained aircraft. Still, even with the best planning and the most cooperative equipment there are a host of en-route challenges that can compromise the safety, efficiency or comfort of flight.

Familiarity may not always breed contempt but it often results in a marked complacency toward those basic elements of good airmanship that are so necessary. Some of those items may seem elementary but only because they are easily dismissed after some cursory attention in early training. Has anyone challenged your techniques for trimming, leaning, descending from altitude, finessing your autopilot, managing a fuel load or organizing your flight bag since basic training? No? Allow me. Please.

Consider the lowly flight bag, alias brain bag, kit, pilot's purse. That miserable millstone is going to dog every hour of every flight throughout your career. It deserves some thought.

There is an old airline story about a very senior captain. Seems that he almost accepted early retirement because he just didn't have the energy to keep up the pace. When he had to carry his flight bag any distance—and just up to the cockpit is a considerable distance in a 747—he noticed an extraordinary amount of fatigue. Finally, nearing the end of his rope, he set about to lighten up that old flight kit by removing any unnecessary items. When he got to the bottom, he discovered four bricks that had been added one by one by some junior type who had hoped to discourage the old-timer into retirement. When the bricks were removed, he felt like a new man.

There seem to be two schools of thought in the great flight bag controversy. There are those who will go to any length to reduce the weight and/or bulk to an absolute minimum. Then there are a few guys who appear to spend their entire career putting things in, without ever taking anything back out. I carry a fair number of accessory items in my kit, and I have found several very useful ones that you might like to try.

Masking tape. There is broad speculation in the airline industry that a shortage of masking tape would spell imminent disaster. Its uses are legendary and range from the mending of flight attendants' skirt hems to the repair of cockpit panels, doors, instruments and air ducts. The stuff is cheap and durable, can be written on with pen or pencil and will stick to almost anything.

Felt-tip pens. I carry about four different colors for marking SIDs, STARs and sometimes the en-route charts. Often there are several procedures on one page, and with a little color coding the ambiguities tend to disappear. I mark the written text as well as the profile and often find that the simple act of tracing and under-lining resolves many potential misunderstandings. Then when I do need a quick reference, I can refer directly to the appropriately colored material.

Road maps. Lots of guys carry complete road atlases. When the major VORs are marked on these maps with a magic marker, you have a quick reference guide to your position and your progress in relation to, say, Yosemite National Park or downtown Chillicothe —which can come in handy with inquisitive passengers on board.

Old en-route charts. The single map I most often use is the US (HI). Since I depend on that one chart so much, I have developed a little system. When a new chart is issued, I begin copying the en-route communications frequencies right on the chart. By the time the next map is issued, my old one is liberally documented with ATC frequencies for ready reference in the event of communica-

tions problems. That old map is then kept as a spare and a catalogue of working frequencies.

A-1 sauce. You know that brown spicy glop that comes in a tall,
skinny bottle? It makes a great cover for those overdone hockey
pucks that the caterers call "filets." It's also good on pork or lamb.
Maybe you would prefer Worcestershire, ketchup or peanut butter.

License and document numbers. Loss of your pilot's license or
passport can be a real hassle. If you note all the serial and/or
identification numbers and store them in the back of your Jepp
manual, along with the phone numbers for the FAA, FCC and the
State Department, the notification process is substantially less
complicated.

Car keys. Ever return from a trip at 0200 and discover in the
parking lot that the keys are in Keokuk? And the other guys have
just left? That's why I carry a spare set in my flight kit for each
family car.

Hand level. The hand level is a pocket-sized device used by
surveyors to estimate lines of equal elevation. At cruising altitudes, a hand level can tell you with considerable accuracy whether
your flight path will clear the thunderstorms ahead. Radar may
show you the cell's intensity, but that hand level will tell you
whether a fuel-consuming climb or deviation is necessary.

Tools. There are a few small tools that are perennially useful.
Naturally, a screwdriver or two and a pocket knife will see lots of
use. I have two other favorites. Most of the knobs on the airplanes
I fly are secured with tiny set screws tapped for either a .050-inch
or a 1/16-inch Allen wrench. Since loose knobs can create really
serious problems, I carry two each of those sizes. My other favorite
is a small vise grip for universal applications from buffet doors to
eyeglasses.

Food. It can be really difficult to eat with any regularity during
a busy flying day, and yet it is essential to keep your own blood
sugar up to normal levels. Candy bars or other sweet snacks will
induce a more serious problem called rebound hypoglycemia. When
you start to sag from lack of nourishment, a quick snack of fruit,
nuts or any high-protein food will go a long way. Check your local
health-food store for appropriate ready-packaged snacks and food
bars.

There are lots of other little items for comfort and safety: aspirin,
toothpicks, plastic raincoat, ring reinforcements for repairing
manual pages and antacid for the hockey pucks. One item I plan

to add is a very small 6X monocular, or telescope, just for the personal enjoyment of sightseeing. If you have a lot of weight and balance or fuel computations, consider a calculator. Actually, there is no end to the possibilities for a personalized flight-kit collection.

One word about the bag itself. Look for very good quality snap-locks, and avoid piano hinges on the flap. They wear holes in your pants legs and prevent the flaps from dropping all the way down flush with the bag's sides when they are out of the way. Good-quality vinyl is cheap, durable and attractive and will outlast the hardware and stitching, at any rate.

Now if you are really senior, go check for those bricks, and if you are junior . . .

Radio Terminology

34

If you are a junior, you may spend much of your en-route time handling basic communication chores. Even if you own the business, however, you will have frequent occasion to talk on the radio and that insignificant task carries some unique risks of its own. Pilots have always communicated in a curious mixture of precision and informality that allows, quite naturally, for misunderstandings. We enjoy a light approach to the dry routine of radio communication—the "Whiskey" ATIS information just naturally becomes the "booze news," and the most conscientious pre-flight routine is never conceded to be more than "kicking the tires."

Even the act of flying itself takes on almost any other name. If you notice, pilots never fly—they "sashay," "slide," "mosey," "proceed," "navigate" or "drive." "Flying" seems to be an act reserved for insects and birds.

I remember, as an eager young co-pilot, asking the captain, "Should we circumnavigate this thunderstorm up ahead?" "Naw, better drive around it," was the sage reply from one better versed in flying than in appropriate terminology. Nevertheless, his instruction was perfectly clear.

Occasionally some simple verbal confusion contributes to disaster, though. Take that old story about the unhappy co-pilot. He reports to operations one morning deeply discouraged by his girlfriend's last letter and the lamentable status of his checking account. During taxi-out the captain tries to encourage him with sympathy and wisdom, but nothing seems to work. As a last gesture, during the takeoff run, he shouts, "Cheer up, Fred," only to feel the airplane mush into the runway as his depressed partner mistakenly raises the gear.

I once witnessed a fantastic misunderstanding in a simulator. The student flight engineer was handling systems emergencies by the book. When the engine fire warning sounded, he correctly remembered the initial four steps:

1. Throttle to idle.
2. Feather button pushed.
3. Mixture idle cut off.
4. Fuel selector off.

Sure enough, starting from left to right he retarded number-one throttle, feathered number-two prop, shut down number-three engine with the mixture and closed the fuel to number four.

Impossible, you say? In the area of human communications, nothing is impossible.

Consider an airline crash west of Dulles several years ago. When the pilot was "cleared for approach," he descended to the initial approach altitude of 1,800 feet almost 30 miles from the airport because he believed that an approach clearance was authorization for an immediate descent. Subsequent public hearings on the crash have uncovered a broad confusion by all parties over standard air-traffic terminology.

An instruction as common and crucial as "cleared for approach" ought to have a specific and precisely defined meaning. Unfortunately, it doesn't. That common clearance is not alone, either—there are dozens of ambiguous communications in use. Some are certainly more critical than others, but here are a few I have encountered recently:

• "Braking action good" (or poor or nil or whatever). This whole category may comprise the single most subjective and ill-defined report that pilots, especially jet pilots, are forced to use. For the past ten years I have made a living in the right seat of various airliners. In that time I have transmitted hundreds of braking-action reports dictated by captains, and any similarity between these reports is mostly coincidental. It's not the captains' fault. There just aren't any satisfactory guidelines.

• "Towering cumulus, all quadrants." That report means anything from active tornados in Tulsa to 9,000-foot tops in San Francisco.

• On final approach in marginal weather, we are obviously a little close to the traffic ahead, so we ask what type it is. "De Havilland," says the tower. Lot of help that is. Now we know that it's either a Twin Otter at 50 knots or a DH-125 at 150 knots. Why not say "airplane"?

• Remember the incident at O'Hare in which a DC-9 clipped a CV-880 on takeoff? That disaster claimed dozens of lives due to a lack of understanding between pilot and controller. A few short weeks after that disaster, when I would have expected the O'Hare controllers to be somewhat subdued, we were vectored out of a holding pattern by ORD approach with the cute clearance, "Turn north and slow to south."

Now that phraseology is clever, adroit, cool and even witty, but it has no place in the control of air traffic in congested airspace. "Turn to a heading of three-six-zero and slow to 180 knots" is fine with me. The stakes are too high to justify adolescent jargon in the air traffic environment.

• You've heard of the two-pound bird whose impact your windshield can sustain at 400 knots—and the 400-pound bird you'd better not hit at more than two knots? I never appreciated that joke until I saw a 400-pound bird. It happened when making an NBD approach to Reykjavik, Iceland, in light rain, with "altimeter setting nine-nine-one." We set it in and began the letdown to 400 feet over the ocean. When we broke out of the overcast, a duck swimming in the ocean up ahead appeared to be a minimum of 400 pounds because he occupied a substantial portion of our total view through the windshield. But our fascination lasted only until we realized that "nine-nine-one" is 991 millibars, which converts to 29.26 inches on our U.S. altimeter. We had been almost 600 feet too low throughout the approach and could easily have descended into the water.

Terminology. As L'il Abner says, it's amazin' and confuzin'. It would help if the FAA could clear up the most obvious problems of ATC language. Maybe it's time for us in the cockpit to lead the way.

Pilot Reports

35

One way we can lead is to take the initiative for those in-flight reports that may be useful to others. Professional flying has become so computerized, organized and standardized we pilots sometimes forget that, except for the traffic situation, more knowledge comes from aloft than can ever be discovered on the ground. We acknowledge and comply with ground-initiated messages and tend to forget there are several vital reports that can, and should, originate from the cockpit.

Several reports fall in the category of general advisories. But these radio messages deal with information of vital interest to others.

Icing reports. FAR 91.125 requires you to report any unexpected weather conditions encountered and any other information relating to flight safety.

In-flight icing often fits both of these criteria. Your report will be of value to other aircraft because only you may know about the existing hazardous condition. But you must make your report on it as accurate as possible. Proper terminology is detailed in the *Airman's Information Manual* (AIM), but it can be summarized as follows:

All ice is either clear (glossy and translucent) or rime (rough, milky and opaque). Besides these two broad characteristics, airframe icing should be classified according to intensity.

• A trace of ice is a minimal accumulation that will not adversely affect even the aircraft without anti-ice/deice equipment unless encountered continuously over one hour.

• Light icing may present a problem if flight is prolonged. Occasional use of deicing/anti-icing equipment will preclude any problems.

185

• Moderate icing will require the use of deice/anti-ice equipment or diversion.

• Severe ice cannot be shed even with the use of deice/anti-ice equipment. An immediate diversion is necessary.

Turbulence reports. Another phenomenon that often triggers an in-flight report as specified in FAR 91.25 is turbulence. Again the AIM is very helpful in defining accurate terminology.

• Light turbulence causes slight erratic changes in altitude or attitude. Light chop causes a slight, rhythmic bumpiness without appreciable changes in altitude or attitude. Passengers would have little or no difficulty in walking in either case.

• Moderate turbulence causes some changes in altitude, attitude and/or airspeed, but the aircraft remains in positive control at all times. Moderate chop causes rapid bumps or jolts without changes in aircraft altitude or attitude. Walking is difficult, and unsecured objects may be dislodged.

• Severe turbulence causes abrupt changes in altitude, attitude and airspeed. The aircraft may be momentarily out of control and walking impossible.

• Extreme turbulence causes the aircraft to be violently tossed about and practically impossible to control. It may even cause structural damage.

Besides these categories of intensity, turbulence or chop should be reported as occasional when it is occurring less than one-third of the time, intermittent when it occurs one-third to two-thirds of the time, and continuous when occurring more than two-thirds of the time.

Navaid reports. FAA encourages us to report on substandard navaid performance and most pilots are quick to yell when a station goes down. There are, however, several atrocious navaids to which the aviation community has simply accommodated itself, and these facilities need to be reported on a continuing basis until they are rendered more serviceable. Sparta VOR, the eastern terminus of J-70, has been so bad for so long that pilots and controllers just take it for granted despite the fact that the radials of J-70 sometimes wiggle perilously close to other east-west airways. The Harrisburg, Pennsylvania, VORTAC has been notoriously unstable for years, and Vance VORTAC down in South Carolina can throw you a wicked curve. If we'd all take those navaid reports seriously, we might eventually enjoy better facilities.

PILOT REPORTS

Migrating bird reports. Migrating birds are a serious hazard to air navigation every year. Brief reports of their location, altitude and direction can be of great value to other pilots and air traffic controllers.

Forest fire reports. Enormous amounts of timber and woodland are destroyed by fire in this country every year. As with all fires, early detection can minimize the damage. I'm not sure how it works, but many FAA traffic facilities can communicate directly with Smokey the Bear. So when you see a telltale wisp of smoke during the fire season, don't be afraid to tell the controller. Your early report could save a lot of lumber or campsites, or both.

CIRVIS reports. A little-known report is outlined in a document entitled "Canadian-United States Communications Instructions for Reporting Vital Intelligence Sightings." The acronym is CIRVIS, pronounced "Survees," and it provides machinery for you to report on bad guys sneaking up on the United States.

Under CIRVIS you would report hostile or unidentified aircraft, missiles, submarines or surface vessels. In addition, you would report subs or aircraft engaged in suspicious activities or observed in a location or on a course that could be interpreted as constituting a threat.

To make a CIRVIS report you broadcast on the normal en-route frequency and begin by clearing the channel with CIRVIS, or PAN, spoken three times. After that opener, you simply identify yourself and describe the object(s), including numbers and category. Don't forget the position, time, altitude and speed, as appropriate. You may have trouble getting the controller's attention because CIRVIS is a neglected program in this age of detente, but it's there to be used when needed.

This brief catalogue of cockpit-initiated radio reports is far from complete. There are many hazards to flight safety that should be reported briefly for the benefit of others down the line.

Certainly no one likes a motor-mouth on the frequency, but it would be a shame if the system became so one-sided as to stifle pertinent and vital reports initiated from aloft.

HF Radio

36

If you fly over water or remote areas you may have occasion to use high-frequency (HF) radio communication equipment. HF was the norm many years ago before the advent of reliable VHF equipment. Recently, HF radios have staged a comeback in general aviation for flights that require long-range communications.

High-frequency radio is very different from very-high-frequency radio, despite the similarity in names. My first substantial cross-country was in a Navy Beech 18, more affectionately known as a Secret Navy Bomber, after its official designation of SNB. That trip from Pensacola to Minot, North Dakota, was pretty routine except for tremendous headwinds that held our groundspeed below 100 knots the entire way, with one leg averaging only 88 knots.

What I remember best are the long conversations with individual radio facilities—ATC centers were not completed then—as we sought a more favorable route or altitude in the face of that strong northwest wind. All of those conversations took place on the old HF common-frequency of 3023.5KC.

High-frequency communications were and are a mixed bag. Nothing else provides the sheer range of voice communications although reception is often disappointing or nonexistent, due to skip or interference.

When the signal leaves an HF transmitter, it splits into two segments. One part remains near the earth's surface and is called a *groundwave*. Broadcast stations rely on the groundwave to carry their signal to distances around 100 miles. A second part of the HF signal heads for outer space but is reflected back by the ionosphere, which acts like a giant mirror. These reflected signals, called *skywaves*, are more common in the upper portion of the HF band used by aviation and at night. That, in fact, is the reason you can often hear distant broadcast stations on your AM radio after dark.

Since the ionosphere's reflective capabilities vary greatly with time and season, it can be helpful to remember the most obvious changes.

• Higher frequencies (10 to 30 MHz) are generally more usable during the day. I use the mental guide, "high freq at high noon."

• Lower frequencies (2 to 10 MHz) are more useful at night although everything is likely to be open on summer nights.

• During winter months there will be a really pronounced shift at sunset, and the low frequencies will tend to have extraordinary strength and range on winter nights.

Fading is an HF phenomenon in which the signal strength increases and decreases at rates from several times per second to once every several minutes. Fading is most pronounced below 6 MHz and may often be eliminated by selecting another frequency. Fading is often very different for any two frequencies, even those very close.

Skip occurs when the skywave signal is reflected over or past a station. If you don't get an immediate response, try repeating the call in a few minutes when your transmitter and/or receiver may be better positioned to take advantage of the skip.

Forecasts of HF signal propagation are given over Station WWV, the National Bureau of Standards time signal broadcasting facility, at 14 minutes after each hour. WWV can be received on the following HF frequencies: 2.5, 5, 10, 15, 20 and 25 MHz. These are short-term forecasts of propagation conditions in the North Atlantic, but are generally applicable to the northern hemisphere. The actual announcements are given in the form of a code which uses one phonetic word and a single digit. The word identifies the radio propagation at the time the forecast is issued (0100, 0700, 1300 and 1900 zulu-time). The numeral indicates the quality expected during the ensuing six hours. The codes have the following meanings:

Whiskey Propagation disturbed
Uniform Propagation unsettled
November Propagation normal
1 Quality useless
2 Quality very poor
3 Quality poor

4 Quality poor-to-fair
5 Quality fair
6 Quality fair-to-good
7 Quality good
8 Quality very good
9 Quality excellent

If, for example, propagation conditions are disturbed and forecast to be fair during the next six hours, the coded forecast announcement would be "whiskey 5."

At the same time, 14 minutes after each hour WWV broadcasts K-index values and solar-flux data. K index is a measure of disturbances in the earth's magnetic field that have a strong effect on HF signal propagation. The K figures broadcast by WWV range from 0 (very quiet) to 9 (extremely disturbed). A K figure of 5, for instance, would mean that the earth's magnetic field was moderately active and likely to degrade HF communications proportionately.

The solar-flux value is a measure of radio activity being received from the sun. This index is not related to an arbitrary scale like K-index but is an actual measure of received energy on some specific frequency. The lowest solar-flux index ever recorded was 64 and readings near 150 are considered high. When the solar-flux index is low, HF communications above about 10 MHz are likely to be impaired.

As the index increases, those higher frequencies become more usable. Solar-flux index is just a rough handle on your maximum usable frequency.

A typical complete announcement at 14 minutes after the hour would sound something like this:

"The radio propagation quality forecast at 1300 is good. Current geomagnetic activity is normal. The coded forecast is November 7. The K-index at 1300 is 2, tending to increase. The 2,800 MHz solar-flux index is 70 units, tending to remain constant."

The National Bureau of Standards handbook providing complete time and frequency information is available from the Government Printing Office, Washington, D.C. 20402. Its order code is SD CAT C13.10:432.

HF communications have come a long way from the old coffee grinders, and it continues to get better and more reliable. If you are expanding into an over-water operation, that long-range radio will be a must for air traffic control, company communications, weather information and time data. If your experience has been limited to domestic VHF communications, find a friend with some solid HF experience and pick his brains before flying that first long trip.

Poor Runway Judgment Can Leave You Short

37

A young newspaper reporter once had the opportunity to interview the famous oil tycoon John D. Rockefeller, whose personal fortune was at that time without parallel. After questioning Mr. Rockefeller at length on his personal and financial life, the interviewer concluded by asking the multimillionaire, "How much money is enough?"

Mr. Rockefeller's response was brief and provocative. "Just a little more," he said, "just a little more."

Pilots spend their entire flying career asking the how much question in three specific areas. How much runway? How much fuel? How much altitude?

How much runway is enough? How much is necessary for a safe takeoff or landing? How much indeed. That basic question underlies the safety of every flight, and has been asked by at least four generations of pilots, beginning at Kill Devil Hill.

The textbook *Principles of Aviation, Volume I* was published by the Curtiss-Wright Flying Service in 1929. It contains almost 350 mimeographed pages that outline everything a pilot needed to know in those simpler times, half a century ago. The introduction comments with amazement on the recent invention of television and progresses to a detailed analysis of the excellent future for aviation based on technology, civil and military requirements, and the recent introduction of consumer credit to finance such major purchases as an airplane.

Runway judgment was largely a guessing game in 1929, but this old Curtiss-Wright handbook had some straightforward advice:

> When satisfied that the machine is headed in the right direction, that there are no other machines in the path, and that there is distance enough to the nearest obstruction so that the airplane will clear, we give her the gun.

In the 53 years since those instructions were written, airplane performance has been inextricably linked with runway length. Pilots can no longer trust their instincts alone when selecting a takeoff or landing runway because solid runway judgment has become a matrix of interrelated variables.

My purpose here is to stimulate your thinking on the crucial question, "How much runway is enough?" Your flying experience will include thousands of takeoffs and landings, and just one lapse in runway judgment will cause an awful disappointment at the very least. You need a perfect record in a very imperfect system. Consider:

• The crew of a fully loaded DC-8 freighter was cleared to taxi to Runway 1R at San Francisco International Airport. During the takeoff in darkness and rain the crew realized that the available runway was just barely sufficient to allow rotation and lift-off, without any margin for deceleration in the event of an abort. At the end of the runway the heavy freighter was barely pulled into the air over San Francisco Bay by the surprised pilots, who suddenly realized that they inadvertently had taken off from the much shorter Runway 1L.

• A B-747 struck several approach-light stanchions after takeoff, causing significant fuselage damage. With some difficulty, the airplane returned for a safe landing. Investigation revealed that the wing flaps had not been positioned properly to accommodate a last-minute runway change.

• A DC-8 freighter touched down right on the numbers and easily decelerated to about 80 knots on the rain- and slush-covered runway when the tower requested an expedited taxi to the far end before turnoff. At about 1,000 feet from the end, when braking was initiated for a reduction to taxi speed, it became obvious that braking action was nil. The big freighter skated off into the mud, a victim of the slick conditions.

• A BAC 1-11 passed over the runway threshold at 184 KIAS— 71 knots above landing reference speed—and landed halfway down the 5,500-foot runway at 163 KIAS (Knots Indicated Airspeed). The crew was unable to stop on the remaining surface and the aircraft came to rest 728 feet beyond the runway end.

Every flight includes at least two runway decisions, takeoff and landing; both are critical to safety. Aircraft handbooks will provide the basic numbers for gross weight, runway length, elevation and temperature. Pilots must be familiar with this information as a standard against which to evaluate actual conditions. Unfortunately, that textbook standard is accurate only under textbook conditions. Any circumstance that changes those idealized conditions will require some common-sense runway judgment.

AVOID TAKEOFF COMPLACENCY

The vast majority of takeoffs are accomplished within a fraction of the available runway. That happy routine can foster a sense of complacency that compromises legal and technical safety margins. Every takeoff needs to be planned for success *and* for failure. You need enough runway and clearway to allow a safe takeoff and obstacle clearance *or* deceleration to a stop from some reasonable rejection speed.

Airplanes certified to Part 25 standards will have accelerate-stop distances built into runway-length information in their manuals. Part 23 aircraft may or may not. Start with the aircraft handbook numbers and then consider the following elements of basic pilot judgment—particularly when the available runway is close to minimum length.

• Runway alignment distance, the distance necessary to maneuver the airplane into takeoff position, amounts to lost runway. If you start the takeoff 200 feet from the beginning of a 4,000 foot strip, you are really using a 3,800 foot runway, a five-percent loss. Since it is not normally possible to commence the takeoff from the first foot of runway, factor your judgment accordingly, and don't give up unnecessary distance. In some cases a rolling start from the adjacent taxiway can compensate for lost alignment distance.

• Runway slope will have a pronounced effect on takeoff distance and must be considered in your runway judgment, even though it is difficult to obtain accurate slope information. You will not find slope data on any Jeppesen charts and only on a few NOS (National Oceanographic Service) charts. Airport managers may or may not have slope data readily available, and tower controllers almost never have it.

Basically you can judge the amount of runway slope in two ways. Some airport diagrams show the elevation of each runway end so that simple arithmetic will reveal the average slope, although that ignores the intervening humps and depressions. If the runway in question is distorted by a prominent bump or depression, modify your runway judgment accordingly.

Humps detract from effective runway distance because the airplane must accelerate uphill and then, in the event of an abort, decelerate downhill. Depressions work in your favor by enhancing acceleration and deceleration if needed. In all cases balance your runway judgment by an available slope information, even if it is only that available by eyesight.

• Runway environment temperature has a significant effect on takeoff distance. That is no secret. But temperature in the runway environment can be as much as 20°C higher than that at the official airport thermometers, which are specifically located to measure free air temperature over grassy or sodded areas. Each additional degree detracts from actual takeoff performance by degrading aerodynamic and engine efficiencies. When planning a takeoff on those sunny days, factor your runway judgment by consulting the next higher temperature increment in the pilot's handbook or even by some arbitrary mental adjustment. Temperature extremes can be as hazardous as lack of distance.

• Runway slickness will have no effect on your takeoff distance but may preclude a safe stop after an abort. Unbelievably, the FARs require no factors in handbooks to compensate for water, ice, snow, or slush on the runway, so it becomes a matter of pure pilot judgment. The problem is particularly insidious in Part 25 airplanes because pilots may be led to believe that slippery runways are considered in the accelerate-stop distance. In fact, they are not.

• A complete lack of acceleration information leaves the pilot with only speed as a measure of distance traveled along the runway. V_1 speeds *assume* normal acceleration. We need some means of checking that assumption in the cockpit.

In 1970, a DC-8 attempted to depart from Anchorage, Alaska, with the brakes locked on an icy runway. By the time the crew recognized that the acceleration was subnormal, it was too late to prevent a crash that killed 47 people.

In response to another takeoff accident, the NTSB recommended "the use of takeoff procedures that will provide flight crews with time and distance references to associate with the acceleration to V_1 speed." Similarly, the Flight Safety Foundation has recommended these so-called "line-speed" checks to verify normal acceleration.

Surely, a simple accelerometer could be made that would provide these essential data.

• Pilot training for aborted takeoffs is totally inadequate. Engine failure drills in simulators or aircraft invariably are conducted with a great excess of runway so that a relatively casual response is sufficient to stop the airplane on the runway. Line pilots are not given the theoretical and practical training they need to conduct a safe takeoff abort at or near runway-limiting weights.

Decision speed calculations and rejected takeoffs deserve some hard, honest scrutiny by the rulemakers. I fervently hope that the FAA will ignore the inevitable pressures from manufacturers and airlines in favor of more realistic certification and operating regulations.

• Recognition time, the allowable delay between an actual failure and the pilot's reaction, is woefully insufficient in the published decision speeds.

During the certification runs that are made to establish V_1 tables, test crews respond to prearranged engine failures in a clinical, flight-test environment. When the final, and best, data are prepared in chart form for the approved handbook, a one-second margin is added to the test crew's reaction time. Some margin.

Just consider the possible influence of crew inexperience, fatigue, adverse weather, cockpit communications or any number of distractions. Now, doesn't it seem more reasonable to assume that reaction time in the real world would be slower by five seconds, on the average?

Simulator tests and accident findings indicate that a five-second margin would be more useful. At a V_1 of 120 knots, that extra four seconds amounts to 800 feet of runway. If the rollout end of the runway terminates in water, steep terrain or an obstacle, a disaster would be likely.

Ironically, in 1965 the FAA proposed adding 600 feet of runway to the V_1 computation to compensate for recognition time. The proposal was withdrawn.

• Nonengine-related failures are not even considered in the takeoff certification process for Part 25 aircraft, which concentrates on engine failures. One 13-year compilation of 30 aborted takeoffs by air carriers lists only seven caused by engine problems. Seven others were caused by brake and tire failures, another serious threat that consistently had been discounted until the unsuccessful DC-10 abort and fire in Los Angeles. More recent NTSB data indicate that tire and brake failures are responsible for 70 to 80 percent of all takeoff aborts.

Other factors also must be taken into account in determining minimum runway length for takeoff. Tire and brake condition will affect abort capability; aerodynamic drag caused by control-

197

surface deflection in a crosswind takeoff will retard normal acceleration; aircraft loading at or near center of gravity limits can extend the takeoff run.

How much runway is necessary for a safe landing? Airplane flight manuals contain handy graphs or charts showing minimum landing distances or maximum landing weights for specific runway conditions. Unfortunately, that predigested material does not show what margin, if any, there is for error, so that pilots often must guess at the cumulative effects of poor braking action, crosswinds, and equipment failure. It is not reasonable to expect a precise resolution to the landing-distance equation for every set of conditions. Still, if you understand the assumptions that are used to construct those flight manual charts, you will be better able to exercise good judgment.

Essentially, there are three major elements that contribute to those final handbook charts for determining runway distance for landing:

- Aircraft performance.
- Certification procedures.
- Operating regulations.

Aircraft performance. Basic aircraft performance is the most important element in establishing safe, minimum runway lengths.

Stable handling qualities, minimum safe airspeeds, brake energy capabilities and ground handling all are vital design elements that contribute to landing and stopping ability. Anyone in the market for a new aircraft will find significant differences in landing performance among available models in any category, and those differences should be weighed as carefully as cruise speed or climb performance. Once the airplane is in service, however, the line pilot will be more concerned about his personal margin for error. And margins can be understood only in light of the certification and operating rules.

Certification procedures. It is not really possible to determine your own safe minimum landing distance without some knowledge of the way your airplane was certificated by the FAA. Basically the agency uses three separate standards:

- Normal category airplanes of 6,000 pounds maximum weight or less are certificated under certain provisions of FAR Part 23. For such aircraft there are no landing-distance requirements, and manufacturers are not required to measure or document necessary runway lengths.

Landing distances published for any aircraft under 6,000 pounds gross weight are based on whatever assumptions the manufacturer chooses to make. Some aircraft handbooks are prepared very carefully, with all associated conditions clearly stated.

Others, particularly those handbooks that were prepared prior to the industry's efforts to standardize the format of operating manuals, are not. You, as the pilot, must understand those assumptions before the raw data can be of significant help to you. If the handbook is not clear, insist on a complete and precise explanation from the manufacturer.

Balked landing performances for these under-6,000 pound airplanes must include at least a 200-fpm climb rate at sea level.

• Normal category airplanes over 6,000 pounds maximum weight are certificated under different and somewhat more stringent provisions of Part 23. The landing distance for these aircraft must be measured and documented from a gate position 50 feet above the landing surface at a speed not less than 1.3 times the stall speed of the landing configuration. There is no margin built into these tests, but the basic profiles are reasonable and repeatable.

Balked landing performance for these heavier normal-category airplanes must be a climb ratio of at least 1:30 at sea level, a gradient of 3.3 percent.

• Transport category airplanes of any weight are certificated under the provisions of FAR Part 25. The landing distance for these aircraft must be determined on a dry, hard-surface runway from that same 50-foot, 1.3 Vso gate. Test data are gathered from approaches in a steady glide to the 50-foot height at no less than 1.3 times the landing-stall speed.

After touchdown, stopping distance is based on the drag from the landing flap setting, fully extended spoilers and maximum wheel braking. Reverse thrust or drag chutes may not be used. Again, there are no significant margins, just a reasonable and useful landing profile.

Balked landing climb gradient for Part 25 airplanes must be at least 3.2 percent at sea level.

Each of these certification procedures in Part 23 and 25 is used to establish the *required landing distance*, the distance needed to land and bring the airplane to a full stop (Figure 1). How you use those basic required landing distances depends on the applicable operating regulations and, of course, simple pilot judgment.

Operating regulations. Operating rules contained in FAR Parts

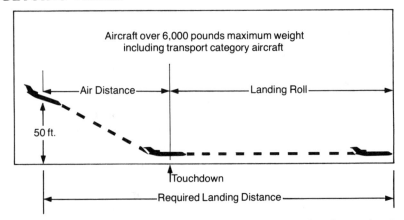

Aircraft over 6,000 pounds maximum weight
including transport category aircraft

Air Distance —————— Landing Roll ——————

50 ft.

↑ Touchdown

————— Required Landing Distance —————

Figure 1. Required landing distance. Test pilots can land and stop the airplane from 50 feet above the threshold at a speed of 1.3 V_{so}. That value is known as the required landing distance. Most general-aviation flight manuals use these bare numbers on their landing-distance charts. There is no margin.

91, 121 and 135 establish the legal requirements for aircraft use. All Part 91 operators and those Part 135 operators using Part 23-certificated aircraft with a capacity of fewer than ten passengers may legally plan landings on runways no longer than the required landing distance demonstrated during certification. There simply is no required margin for error.

Part 121 and other Part 135 operators must add a safety factor to the minimum landing distance. Simply stated, the maximum landing weight must be limited so that the *required landing distance* will not exceed 60 percent of the runway's *effective landing length.* In other words, the amount of runway required by test pilots in the certification process must not exceed 60 percent of the runway available; therefore, the line pilot always keeps 40 percent of the runway as a safety margin (Figure 2). In addition, 15 percent is added to the required landing distance for wet runways. Generally, 40 percent of the available runway should be left to cover the difference between your situation and that of the test pilot.

If your flight manual does not include that cushion, you should. Where necessary, just multiply the handbook landing distances by 1.67 and always insist on that minimum amount of runway (Figure 3). For aircraft weighing less than 6,000 pounds, you may have to dig carefully to establish the basis for handbook numbers.

In all cases, however, strive to establish end figures that will maintain 40 percent of the available runway as a margin. It may

POOR RUNWAY JUDGMENT

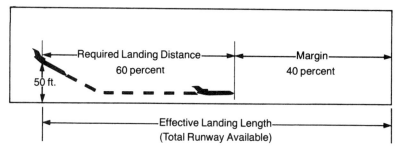

Figure 2. Part 121 safety margin. Air carrier and Part 135.2 operations are predicated on maintaining 40 percent of the runway length available as a safety margin. It's a good idea to base all flight operations on that intelligent margin.

sound like a comfortable cushion, and under good conditions it is, but consider the elements that can consume some or all of that 40 percent.

FACTORS AFFECTING AIR DISTANCE

Any landing distance assumes precise airspeed and glidepath profiles from the 50-foot gate. Excess altitude will extend the distance proportionately and low approaches increase the risk of an undershoot on a landing.

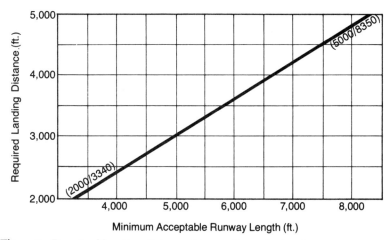

Figure 3. Required landing distance × 1.67. Enter the landing-distance chart from the flight handbook and obtain the minimum acceptable runway length that allows a 40 percent margin.

201

If 50 feet over the threshold aims you 1,000 feet down the runway, then 100 feet will move the aiming point to 2,000 feet, assuming your glide angle has not changed. When you are high, there are three options:

• Continue a stabilized approach and accept the fact that you are going to land long.

• Steepen the approach, realizing that you will need added airspeed or power at the bottom to flare, and accept the fact that you still are going to land somewhat long and maybe hard.

• Go around.

Flat approaches—below the 50-foot gate—can get you onto the runway a little early, perhaps as much as several hundred feet, but they greatly increase the risk of an undershoot, which could be far more disastrous than rolling off the far end at reduced speed. Also, low approaches usually are flatter than normal, thus they often result in long, rather than short, landings.

Low approaches are rationalized by saying that the runway in use was only as long as the handbook said was required, so the pilot felt it necessary to land on the very beginning. That lame excuse reflects both poor judgment and a lack of understanding. These pilots are confusing landing roll with landing distance. The FARs mandate that landing distances—at least for all aircraft over 6,000 pounds—must include the air distance from 50 feet over the runway end. In any case, if you feel the runway is so short that you need to cross the threshold below 50 feet, seriously consider another runway or even another airport.

Certification results, and thus handbook landing distances, also are based on a speed of 1.3 times the stall speed. Pilots of turbine-powered aircraft call it reference speed or V_{REF}. If you arrive at the 50-foot gate at some speed above V_{REF} you again have three options:

• Continue the approach and land fast. You will use about two percent more runway for each one percent of excess speed.

• Continue the approach and bleed off the excess airspeed before touchdown. This will require about 700 to 1,000 feet of runway for every five knots of excess speed and is by far the poorer technique.

• Go around.

When you are fast at the 50-foot gate, a larger than normal power reduction will help, but you still will be ahead of the desired profile even if the speed is back to normal at touchdown, and that alone will reduce your 40-percent margin.

There are at least two other factors that affect air distance when planning a landing. First, any attempt to flare very close to the runway for a smooth touchdown will extend the total landing distance beyond what is assumed in the handbook charts. Second, approaches and flares at high density-altitudes are begun at a disadvantage because certification results are not corrected for temperature effects on these phases. When temperatures are above standard, true airspeeds will be higher and flare distances longer. If you do notice temperature compensations in your manual, they probably apply only to the ground roll and the balked-landing climb gradient.

VARIATIONS IN GROUND ROLL

Stopping distance on the ground is predicated on the rapid and correct use of all deceleration devices, including maximum braking within seconds after touchdown. No matter how good the approach and touchdown are, you will need to keep the pressure on throughout the landing roll if you hope to come close to certification results. Otherwise you will sacrifice some portion of that 40-percent safety margin.

On dry, smooth runways, wheel-brake potential is probably much greater than most of us realize. By definition, maximum braking is obtained when wheel rotation is sustained just above a skid. Even with the most sophisticated anti-skid device, the average pilot may never use more than 70 percent of the braking power available to him. But anything less than aggressive, forceful use of the brakes, right to a complete stop, will reduce your 40-percent margin, and that is on a smooth, dry runway.

When the concrete is wet or icy, ground roll increases by measurable, and sometimes dramatic, amounts. Normal braking coefficients at typical landing speeds on dry concrete are about .40. Figure 4 shows how those optimum coefficients are reduced by dampness, water and slush. Notice also that maximum braking is always attained at slower speeds and that the last 50 percent of speed decay normally will occur in much less time and distance than the first 50 percent.

Braking action reports can be of significant help in evaluating landing conditions, even though they are subjective and transitory. It would be helpful if there were a recognized standard for these reports but, so far, there isn't. Braking action can be measured in any way that the airport manager deems acceptable and some methods are downright questionable.

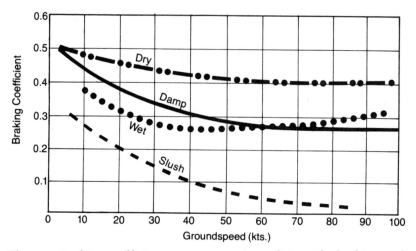

Figure 4. Braking coefficient versus runway condition. The braking coefficient is affected by the condition of the runway. Dampness, water or slush will reduce braking efficiency considerably; these conditions must be weighed.

One news report stated that the manager of a major Alaskan airport has used a pickup truck and a lunchpail for 25 years to measure braking action. He puts the lunchpail in the floor and accelerates down the runway. If the pail tips over when he hits the brakes, braking action is reported as good. If the pail wobbles around, braking is judged fair. If it only wiggles a little, the braking is reported as poor.

If that method sounds frivolous, don't be too quick to dismiss it. At least that airport has a consistent procedure. Many airports have little or no consistency.

Some airports are using a device known as a Mumeter, and others are using a device called a James Brake Decelerometer (DCM). Many use only subjective pilot reports of perceived braking action.

The Mumeter is contained in a trailer, which is towed at a constant speed behind any car or truck. After a single, continuous run down the runway, the Mumeter will have generated a tape readout of runway slickness which is usually reported on the NOTAMs for three or four equally spaced points on the runway. A NOTAM reading "Mumeter .40—.55.—40," for instance, would indicate that braking was fair at the approach end, good in the middle and fair at the far end (Figure 5).

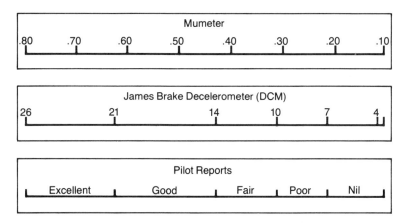

Figure 5. Braking action equivalents. Mumeter and DCM readings, and pilot reports, are all possible souces about the braking action on a given runway. It is wise to obtain at least one of the reports before landing.

Mumeters are accurate, but they are expensive and extremely delicate. Few airports have them, and even fewer can keep them properly calibrated. But you may see an occasional NOTAM that mentions Mumeter readings, so you may want to copy Figure 5, which shows the Mumeter and DCM scales, and keep it.

The James Brake Decelerometer is a suitcase-size device that rides on the seat of any available car or truck. DCM readings are made by accelerating the vehicle to 40 mph, braking hard and noting the reading on a larger meter. DCM values also may be found on some NOTAMs, in the same format as Mumeter readings. A NOTAM of DCM 12-12-8, for instance, would indicate fair braking action at the approach end and in the middle, but only poor braking at the rollout end.

Pilot reports of braking action are limited to five adjectives. Although these simple reports, which are disseminated by Automatic Terminal Information Service (ATIS) broadcasts and tower controllers, are subjective, they are intended to convey the following meanings:

Excellent braking Maximum energy stops are possible.

Good More braking is available than will be used in an average deceleration. If a maximum-energy stop were attempted, some distance in excess of certified stopping distance would be expected.

Fair Sufficient braking and cornering are available for a well-flown approach and landing using light braking. An over-speed or long touchdown would result in an extremely low safety factor, even considering the 40-percent runway margin. Careful planning and good judgment are required.

Poor Very careful planning, judgment and execution are absolutely essential. The crosswind becomes an essential consideration. There is little room for error, even considering the 40-percent runway margin. Care must be exercised in every facet of the operation and careful evaluation of all existing conditions is necessary.

Nil Extremely slippery, with poor directional control even at taxi speeds. A 40-percent runway margin would not be enough.

Most airplanes must rely on wheel brakes as their sole means of stopping. But even those aircraft equipped with other deceleration devices will need to assess the braking action carefully. Anything less than smooth and dry conditions will require a considerable amount of that margin, but only experience can tell how much.

Remember that in all cases maximum braking will be achieved by sustaining the wheels just above a skid. On a dry surface this will require heavy pedal pressure, which normally will not be necessary with the 40-percent cushion.

On a slick runway your instincts may demand heavy brake pressure in the face of marginal stopping conditions, but that could lock the wheels or induce hydroplaning. Hydroplaning is not new and it certainly is not limited to high-performance or turbine-powered aircraft. One pilot friend of mine, for instance, encountered serious hydroplaning in a Cessna 182. Jets and turboprops are especially susceptible due to their higher groundspeeds, but nearly any road vehicle or wheeled aircraft is capable of hydroplaning when the conditions are right.

It's time to review the problem and how to cope with it. Hydroplaning is like aerodynamic stall in that it involves a loss of control—usually at high speed. Remembering that relationship provides an incentive to understand and deal with the hydroplaning problem.

Technically, hydroplaning is a loose assortment of three different but related problems. The only common element is water.

Dynamic hydroplaning results from a combination of high groundspeeds and flooded runway surfaces. As the airplane accelerates for takeoff, partial hydroplaning begins as the moving tires contact and displace the stationary water. At moderate speeds a fluid wedge forms and partially separates the tires from the

206

runway. At some higher speed the hydrodynamic force developed under the tires equals the aircraft weight so that the tires are lifted completely off the runway surface. In this extreme condition all tire-to-surface friction is lost, and the vertical component of that fluid wedge produces a spin-down movement, which slows and eventually stops wheel rotation.

The critical speed at which this total separation occurs is called the tire hydroplaning speed, or V_p. Partial hydroplaning (partial loss of control) occurs at speeds below V_p, and total hydroplaning (total loss of control) occurs at speeds above V_p, when there is enough water. Minimum critical water depths vary from 0.1 to 0.4 inches, depending on tire and surface conditions. In reality, any reports of standing water on the runway should alert you to the possibility of dynamic hydroplaning.

Tire condition and particularly tire pressure will significantly influence the hydroplaning potential. The tire hydroplaning speed (V_p) may be closely approximated for any given vehicle by the formula $V_p = 9 \sqrt{P}$ where P equals tire pressure. If, for instance, your airplane's tires are inflated to 100 pounds psi their $V_p = 9 \sqrt{100}$, or 90 knots.

Viscous hydroplaning can occur with very slight amounts of moisture and at groundspeeds well below V_p. The major difference is that viscous hydroplaning requires an extremely smooth runway surface.

Water on newly surfaced asphalt runways, or on touchdown areas with heavy coatings of rubber from repeated landings, forms a tenacious film that can completely separate tire from pavement at speeds 35 percent below V_p. When that film is present, total separation, and thus total loss of control, can be produced by a water thickness of only .001 inches—about what you might expect from morning dew.

So expect viscous hydroplaning on unusually smooth runway areas, and near the extreme ends of any busy runway where rubber deposits are visible.

Steam, or reverted rubber hydroplaning, requires some moisture in conjunction with a locked-wheel condition. It may occur at low speeds, such as during taxiing, and can follow dynamic or viscous hydroplaning.

When runway surfaces are wet or even damp, the heat from a locked wheel can produce steam in the tire footprint area, which may revert the rubber to its gummy, uncured state. This tacky, reverted rubber creates an excellent seal to enclose the footprint area and entrap the steam, which then superheats at temperatures up to 260°C, causing the tire to lift completely off the pavement due to the steam's pressure.

Hydroplaning is, indeed, a triple threat, but you can minimize your exposure to it with good maintenance and proper piloting technique.

Tire pressures should be carefully maintained at the manufacturer's recommended values. Since V_p is related to tire pressure it might seem logical to increase that pressure and thereby raise the V_p for your airplane. In fact, that logic is sound although any increase in the tire pressure will detract from dry runway braking conditions. Recommended pressures are a compromise between braking coefficients and hydroplane speed; don't make arbitrary increases, but do check tires for correct inflation.

Deep radial ribs provide excellent protection against hydroplaning, so monitor tire wear carefully and don't fly with worn treads.

Good pilot technique can substantially reduce your exposure to all types of hydroplaning if you observe a few cautions:

• Know your airplane's V_p and be aware that any groundspeed above V_p on a wet runway will present a risk of hydroplaning. When landing under conditions conducive to hydroplaning make the final approach at the minimum safe speed and touch down firmly to break through the fluid film.

• After touchdown use the spoilers early, if possible, to transfer the aircraft weight to the wheels. Also, ground the nosewheel quickly for maximum directional control.

• Use early and heavy reverse thrust to decelerate the airplane below V_p. Remember, however, that heavy reversing on a slick runway can cause directional problems.

• Brake cautiously and be prepared to ease pedal pressures as soon as you detect a locked wheel condition, then reapply brakes judiciously when wheel rotation resumes.

The best braking results will occur if the pilot begins cautiously and then progressively increases pedal pressure to stay just above a skid as the airplane slows.

USING OTHER DECELERATION DEVICES

Airplanes equipped with ground spoilers derive two individual benefits when those wing-mounted panels are deployed:

• Total aerodynamic drag is increased, which increases deceleration slightly, particularly at higher speeds.

• The coefficient of lift is reduced abruptly, so that the airplane's weight is transferred from wings to wheels for increased wheel brake effectiveness.

When installed, spoilers provide easy and crisp response with little or no risk of directional control problems, even on wet and slippery runways. They should be on the landing checklist and used for every landing.

Reverse-thrust and drag chutes are the only margin that you can buy. They are not taken into account in any FAA certification or operation regulations, so their use will provide some measurable reduction of the required landing distances that were calculated during certification tests.

There are significant differences in the mechanics of jet and prop reverse, but their applications are very similar. Any reverse, or beta range or ground fine prop position is most effective at high speed, so it should be used as early as possible during the ground roll. On anything other than a dry runway, the initial application of reverse thrust should be cautious. Reverse on a slick runway can cause serious directional control problems, particularly in a crosswind (Figure 6).

On a slick, crosswind runway, any airplane will tend to weathercock and drift toward the downwind edge because the tires have little traction and the airframe presents a large sail area to the wind.

When reverse is applied in this situation, the thrust vectors have a side component that drives the airplane farther toward the runway edge. If the pilot acts instinctively and yaws the airplane back toward the centerline, he increases the thrust's side component and accelerates toward the downwind edge.

The proper corrective action is to reduce reverse thrust, or even return to forward thrust, long enough to regain directional control. When the airplane is back on the centerline and tracking straight, you can cautiously try reverse again.

Drag chutes can significantly reduce required landing distances below FAA certification values by increasing aerodynamic drag. When properly installed and operated, a drag chute expands your 40-percent margin, but it also can be a problem on slick, crosswind runways.

If the crosswind is strong enough, the parachute simply will pull the airplane sideways or yaw it around. When directional control

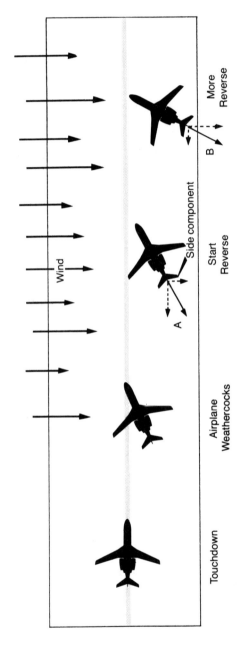

Figure 6. Crosswind on a slick runway. If the airplane weathercocks and drifts downwind, turning toward the centerline and applying more reverse thrust only aggravates the problem.

becomes critical, the drag chute must be disengaged from the aircraft.

Drag chutes and reverse thrust are real bonuses, but, ironically, they create the biggest problems when they are needed most. When the runway is slick and short, use reverse or the drag chute at the earliest possible moment and stay alert for directional control problems.

One elementary note on the use of ground spoilers, reverse and drag chutes: don't use them in the air. You might expect that people would know better, but accident records continue to show an occasional purposeful in-flight deployment, often on high, short final. Usually, the results are disastrous.

HOW MUCH IS ENOUGH?

How much runway is enough for a safe landing? Only you, the pilot, can answer that question provided that you have a clear knowledge of the applicable regulations and operating conditions.

As a minimum you should know the following rules:

• Landing information for airplanes under 6,000 pounds is completely at the discretion of the manufacturer. You must determine the associated conditions or assumptions:

• Aircraft weighing over 6,000 pounds and operated under Part 91 are likely to have landing information that shows only the actual amount of runway needed during certification tests.

Multiply these required landing distances by 1.67 to obtain the desired 40-percent runway margin.

• Part 25 aircraft may be operated under Part 91 one day and 135.2 the next. Even if you are brave enough to operate on the absolute, legal minimums, those minimums will increase by a factor of 0.67 when moving up to the 135.2 operation. Flight manuals may not make this clear and you may have to do your own arithmetic (Figure 3).

When you have taken into account the manufacturer's assumptions and assured yourself of a 40-percent buffer, resist the temptation to relax, particularly on wet or slippery runways. Even 40 percent may not be enough when problems are compounded. With that in mind, consider the following, especially when landing close to the limits:

• Strive to hit the 50-foot, V_{REF} window right on altitude and airspeed in a stabilized condition. Each extra foot of altitude will cost about 20 feet of runway. Each one percent of additional airspeed will require at least two percent extra stopping distance.

• Sacrifice the smooth touchdown for a firm, solid landing.

• Raise the spoilers and ground the aircraft's nosewheel as soon as you possibly can.

• Once stabilized, use reverse cautiously at first and then increase to maximum allowable. Be alert for directional control problems on slick runways with a crosswind.

• Use enough brake pressure to keep just above a skid. With anti-skid brakes, use enough pedal pressure to cycle the anti-skid every two to three seconds.

When there is water on the runway, delay braking until the airplane has slowed below hydroplane speed—nine times the square root of the tire pressure.

• Reduce speed to a crawl before attempting to turn off a slippery runway. Tires have less cornering ability when brakes are being used. Hydroplaning, once begun, can continue down to taxi speeds, with serious loss of cornering ability.

• If you are unable to stop, sliding straight off the far end is preferable to any sideways excursion, unless the runway happens to terminate in rough terrain or deep water. At the far end your speed will be minimal. Also, landing gear struts are not engineered for excessive side loads.

Flight handbook figures for landing distances are merely a reference point from which to begin the landing distance equation.

Each individual must assess those values in light of certification procedures, operating regulations, actual conditions and personal proficiency. When each element is evaluated, you will have a valid estimate of how much runway is enough.

Fuel Consciousness

38

Actually, there's not much to this aviation business if you can just pin down a few basic items:

- Never land or takeoff without enough runway.
- Don't run out of gas.
- Don't fly into the ground.

Unfortunately, we often get wrapped up in details and forget the important stuff, like how much fuel is enough.

Remember that old story about the young pilot who dead-sticked his trainer into a farmer's field? When the ancient and grizzled instructor arrived to fly the airplane out, this bright student explained his error in two words.

"Fuel starvation," he said, with an embarrassed look.

"Nah," responded the instructor. "You just ran out of gas."

Surprisingly, professional pilots are still running out of gas at an alarming rate and for a variety of reasons.

- The 44-year-old captain of a twin turboprop commuter ignored his chief pilot's suggestion to load more fuel and took off for New York's Kennedy Airport without even an adequate weather briefing. Approximately 2 hours and 16 minutes later, after two missed approaches, both engines flamed out due to fuel exhaustion and the aircraft ditched in 60 feet of water. There were no survivors. The NTSB determined that "The probable cause of this accident was fuel exhaustion resulting from inadequate flight preparation and erroneous in-flight decisions by the pilot in command."

- A DC-8-61, operated by a major U.S. trunk carrier, arrived at its destination airport with approximately 65 minutes of fuel remaining. When the landing gear lever was placed in the gear-down position on final approach, the two main-gear lights would not indicate down, and the crew entered a holding pattern to try to fix the problem. After one hour of holding, the airliner consumed its last pounds of precious fuel and the crew dead-sticked the airplane into a wooded, populated area.

The NTSB determined that: "The probable cause of the accident was the failure of the captain to monitor properly the aircraft's fuel state and to respond properly to the low fuel state and the crewmember's advisories regarding fuel state.

"Contributing to the accident was the failure of the other two crewmembers either to fully comprehend the criticality of the fuel state or to successfully communicate their concern to the captain."

• A U.S. flag-carrier 747 arrived at its destination with 1 + 20 worth of fuel on board. After 56 minutes of holding and vectoring, destination weather fell below minimums, and the flight diverted to its planned alternate 18 miles away. The radar-vectored diversion covered 78 ground miles and terminated in an ILS approach. On the runway, engines one and four flamed out because of fuel starvation. The airplane, with 243 people on board, had to be towed to a gate.

Fuel consciousness is an essential element of good airmanship and every pilot who has ever soloed knows it. Consider the master fuel plan of one famous aviator:

> I have divided my reserves. The first I class as reserves for success; the second, as reserves for failure. I depend on my reserves for success to land me on the aerodrome at Le Bourget. I depend on my reserves for failure to let me live if I can't get through to Paris, and, if possible, to save my plane.
>
> The ability to turn back is my greatest reserve for failure. If weather becomes too thick, if I encounter headwinds, or if some fluctuating gauge or engine roughness indicates danger, I can turn back and start the flight again. Here, too, the amount of fuel I carry may be of top importance.
>
> Extra fuel is my greatest reserve for success. With it I can ride through night and detour storms. And the long coastline of Europe—I can be hundreds of miles off course when I strike it and still reach Paris.

That basic philosophy of fuel planning, formulated by Charles Lindbergh in anticipation of his epic Atlantic crossing, is as valid for all levels of aviation today—and for the foreseeable future—as it was in 1927. No single element is more critical to the safe operation of an aircraft than an uninterrupted supply of fuel.

FUEL CONSCIOUSNESS

Incredibly, that basic truth seems to have lost some of its impact with the advent of air-conditioned, coat-and-tie airplanes.

Fuel consciousness is really the sum of three separate elements: fuel planning, fuel use, and fuel-system management.

PLAN THOUGHTFULLY

The Federal Aviation Regulations are very clear in specifying minimum fuel reserves for flight planning. Subtle differences distinguish the FAR Parts 91, 121 and 135 fuel requirements, but the rules state essentially:

VFR Sufficient fuel to fly to the intended destination plus 30 minutes reserve in daytime and 45 minutes at night.

IFR Sufficient fuel to fly to the intended destination, then from the destination to the alternate, plus 45 minutes. (In some cases an alternate is not required. Reserves for overwater flights are somewhat different.)

The above are, of course, *minimum* legal fuel supplies, although they have become the operational standard for many operators and an absolute maximum for some.

At least two elements of the applicable FARs need to be emphasized. First, minimum required fuel reserves have proven to be properly labeled. Thirty minutes' VFR and 45 minutes' IFR is indeed an *absolute minimum* of protection to compensate for emergencies, navigational errors, ATC deviations or unforeseen weather problems.

Second, the FARs stipulate the selection of, and fueling for, an alternate airport. The alternate requirement may be waived if the destination is served by an instrument approach and the weather is forecast to be favorable, which means a ceiling of 2,000 feet and visibility of three miles forecast for the period one hour before to one hour after your ETA. Nevertheless, the emphasis is in favor of filing an alternate and properly so.

Somehow the aviation community has reversed that priority, so that alternate airports normally are listed only when required by regulation, despite the fact that many operational considerations, aside from terminal ceilings and visibility, might suggest planning and fueling for an alternate. One-runway airports, thunderstorm forecasts, limiting crosswind components and poor braking-action reports all are valid reasons for alternate-airport planning. Major airlines routinely list multiple alternates when destination problems are compounded.

One more note about alternate fuel. Plan enough for a circuitous routing from the destination to the alternate, particularly if they are close together. That fuel-starved 747 carried enough alternate fuel for a straight-line diversion of 18 miles but actually was vectored over four times as far, a major element in its low-fuel arrival.

Fuel planning, then, is the first step in professional fuel consciousness. Lindbergh could not plan more specifically than to indulge his instincts for maximum possible range and endurance. Pilots today have several planning tools available, however, and they all should be used carefully.

• Weather reports and forecasts have become impressively reliable, so that en route and destination conditions are seldom a surprise. This is particularly true where Flight Service Stations are available for weather updates en route.

• NOTAMs advise of current restrictions to airport facilities.

• Flight manuals specify average fuel burnouts that are at least close to reality.

• Fuel-quantity indicators are reasonably accurate, at least in most turbine-powered aircraft. In addition, highly accurate fuel totalizers are now available as an option on a number of aircraft. (But beware of fuel gauges on any aircraft, particularly smaller recip models. The clock is often your best fuel gauge, even in the most sophisticated aircraft.)

Unfortunately, the very precision these planning aids provide may seduce the pilot away from genuine fuel consciousness. Most flights are planned with definite allotments of fuel for burnout, alternate, reserve and holding, when appropriate. Still, once airborne, that total fuel load translates into some measurable number of minutes and miles, to be used in the safest way possible. The confidence resulting from fuel planning must be balanced against the reality of circumstances.

Careful fuel planning is essential to safe flight, but skillful fuel use is even more important and may be very different from the original plan.

FUEL CONSERVATION

Fuel conservation—which is just one element of solid fuel consciousness—is a blend of good planning and careful utilization.

FUEL CONSCIOUSNESS

Real conservation begins on the ground, and several things you can do before takeoff will reduce consumption:

• Carefully select the minimum safe fuel load. Carrying fuel uses fuel, so above all, don't tanker it, except in the most urgent cases. Also, look for closer alternates, select low-traffic-level airports for refueling stops and consider using computer-assisted flight plans. Remember, too, weight reduction in jets allows operation at higher, more fuel-efficient flight levels.

• Load passengers, baggage and fuel for maximum aft CG. As CG moves aft, tail down-forces are minimized, which reduces the airplane's *apparent* weight in the air. Aft CG loading is a proven principle used by all major airlines. You can save perhaps 0.5 percent.

In turbine-powered airplanes, use reduced power for takeoff if it is approved for your airplane (only a few jet models have such approval). If not, nag your manufacturer for approval. Reduced-power takeoffs save fuel and engines, which in turn saves more fuel.

In the air you need to think in terms of specific range, the distance traveled per pound of fuel burned. Fuel flows, tailwinds, groundspeeds and altitudes—each is meaningless alone. You must think in terms of fuel consumed per mile traveled over the ground. You've probably been doing that for years in your automobile and now it's time to apply the same logic to flying.

Each flight should be considered as a whole, with minimum total burnout accorded a very high priority. Within any flight you will find three major opportunities to conserve fuel: climb, cruise and descent.

Propeller-powered aircraft cannot realize any significant fuel conservation by altering their climb procedures, so other considerations, such as passenger comfort and cockpit visibility, are more important.

Jets often can realize measurable savings by reducing climb speeds for higher rates of climb. Manufacturer-suggested climb speeds often are compromises that sacrifice fuel for shorter trip times. If you climb at a slower airspeed—something close to the best-rate-of-climb speed—you will save fuel at the expense of a few minutes' flying time. You also will penetrate the congested lower airspace more quickly for some increased safety. At higher altitudes, always cruise-climb when possible in lieu of making large step-climbs. If you must step-climb, level off 1,000 to 2,000 feet above optimum altitude so that fuel burnout will result in the aircraft's flying at its most efficient height.

Fuel conservation techniques in piston aircraft are often hampered by a lack of precise instrumentation. One useful technique is to reduce manifold pressure about one-quarter inch per hour to maintain a constant true airspeed (TAS). Also keep the CG aft, and in a tailwind reduce cruising speed somewhat.

Other than those simple suggestions the single most important procedure for piston-engine fuel conservation is proper leaning. I was a reluctant student, myself.

TRIM

The second airplane I ever flew was a Navy T-28. That big, round-motored trainer was a mixed bag of noise, vibration, exhaust leaks and oil smears, but it offered performance, stability and fun.

On a summer day in Pensacola, Florida, one hour in that North American Trojan with flight suit, Mae West, parachute, oxygen mask, helmet and gloves was enough to melt any John Wayne type down to a pygmy. But at 15,000 feet, above the southeastern haze level, you could pirouette in three dimensions and forget everything except the slow, rolling blur of section lines and sky, swamp and clouds, dirt roads and sun. It was a very special time in a very special airplane.

I didn't have much trouble flying the T-28 with one notable exception: I could never remember to enrich the mixture during descent. The few times I did remember, I would find that I had never leaned the engine properly in the first place.

Finally, in desperation, my instructor took my wife aside at a party and enlisted her help. After describing the problem to her in appropriate feminist terms, he asked her to whisper a sweet message in my ear each night and morning. That simple message was just two words: "Mixture rich." She did just that for several months and I've never forgotten it.

The art and science of leaning piston engines is one of the better-kept secrets of aviation despite repeated emphasis in magazine articles and owners' manuals. There seems to be a collective mental block that prevents most pilots from ever developing real knowledge and skill in fuel-mixture management. It can be especially infuriating if you fly a wide assortment of piston-engine airplanes, but proper mixture management should be easy for those who confine their activities to one or two of these aircraft. The principles involved are common to all reciprocating engines.

Air is roughly 21 percent oxygen and 78 percent nitrogen, with a smattering of other inert gases. When fuel is burned with air in the cylinders, the oxygen and gasoline react to create the heat of combustion. This heat is absorbed by the inert and created gases, which expand and push on the piston. Mixture is critical to the efficiency of this heat-expansion process.

When a chemist burns gasoline and air in a laboratory environment, he finds that the ideal mixture is an air-to-fuel ratio of 15:1 by weight. In other words, with a mixture of one pound of fuel to 15 pounds of air, the combustion process will exactly consume all fuel and all oxygen. This so-called stoichiometric mixture releases the greatest possible amount of heat energy and produces the highest combustive temperatures. But, surprisingly, this 15:1 mixture is of little use to engines.

As the mixture is enriched from 15:1 in the cylinders of an engine, the combustion temperature is reduced, but the power increases progressively to a ratio of about 12:1. This interesting contradiction (more power with less heat) takes place for two reasons. The total heat released by combustion decreases only slightly within this mixture range (15:1 to 12:1), but the vapor formed by excess fuel adds to the total mass of the expanding charge in the cylinder.

That extra fuel vapor accelerates the burning process so that the fuel's available energy is released more rapidly. Thus the 12:1 mixture ratio provides optimum power-production conditions of heat, mass and flame propagation and is referred to as the *best power* mixture ratio.

The *best economy* fuel/air ratio, however, is obtained by setting the mixture for the stoichiometric ratio (the highest obtainable exhaust gas temperature (EGT), and then *leaning* to a prescribed amount. This leaning reduces temperature, power and fuel flow until, at a ratio of approximately 16:1 to 17:1, the piston engine is producing the most horsepower per unit of fuel. Big radial engines were often leaned to this setting for their maximum efficiency.

In either case, best power or best economy, you must start from the stoichiometric mixture point and that takes accurate cylinder heat temperature (CHT) or EGT information. Naturally, EGT is more sensitive and can allow you to check each cylinder separately —a very desirable procedure, since individual EGT temperatures may vary widely due to uneven intake distribution.

Most horizontally opposed engines are leaned to the stoichiometric ratio and then enriched to best power by reference to some prescribed temperature drop. It sounds counter-productive to ignore best economy in smaller engines, but other considerations such as mass charge, valve wear, and lead fouling often take precedence over theory.

If you fly a piston-engine airplane and want to operate for best economy commensurate with long engine life, there are several points to consider:

• It is not possible to efficiently lean an engine without good EGT information. The stoichiometric mixture is always at max EGT and is the only possible reference point for accurate leaning. If you don't have an EGT now, seriously consider an installation that allows individual monitoring of each cylinder.

• The engine manufacturer's recommendations provide an excellent point of departure, so read everything you can find written about your particular engine and airplane. Write to all the manufacturers involved and conscientiously develop a precise leaning procedure. Even if you think you know how, review your ideas for basic soundness. Bear in mind, however, that for some mystical reason the experts often lean deeper than their own recommendations.

• Remember that leaning can be just as important for very-low-power operations, such as taxiing and descent, as it is at cruise because the worst lead fouling occurs at those low temperatures.

Really excellent mixture control is a fine art, requiring much more information than you are likely to pick up in the airport coffee shop. On the other hand, good mixture control is a satisfying skill that can save both fuel and engines.

One basic note on fuel conservation in any type of airplane:

You must trim the airplane for zero control forces in all three axes, in order to achieve maximum fuel efficiency. The funny thing is that trimming is a sensitive subject with most pilots. Every profession has its sacrosanct subjects. I mean, you just don't tell a doctor how to conduct a physical. Journalists don't readily solicit help with their grammar and style (although some of us could probably use it), and I've never found a plumber who appreciates my clever thoughts on soldering. Each of those subjects is so basic to the profession involved they become individualized and personal.

Aviators have their own catalogue of sensitive topics and none is more sensitive than the subject of trim. Trim is just so basic, so elementary, so much a matter of individual technique, that the subject never seems to arise in the normal course of aviation shoptalk.

Even if you think there is nothing more to learn about trim I urge you to read the next few paragraphs. It is not always a simple subject. And, if you feel self-conscious, read this section in the rest room.

Precisely because no one talks about trim, it tends to be a subtle area of misunderstanding, particularly in turbine-powered aircraft where the penalties of mistrim are greatest. I'm certainly no expert, but I have made a concerted effort for the past year to evaluate and mentally critique trim procedures, starting with my own. In the process I have come to believe there are many areas of misunderstanding on this very basic subject, such as:

- The airplane is in trim when the ball is centered.
- The airplane is in trim when the spirit level is centered.
- The airplane is in trim when the wheel is centered.
- The airplane is in trim when the trim indicators are centered.

Basically an airplane is in trim when all the control forces are reduced to zero and the flight path is constant. Bubbles and balls and such are merely indicators that may or may not be accurate in telling you that the desired zero control forces have been achieved. Naturally, the aircraft must be in trim about three separate axes and the order of trim should start with pitch.

Pitch trim is simple because there are no indicators to introduce doubt. Just roll in enough correction to neutralize all elevator forces for the desired attitude and/or airspeed and the job is done.

Yaw trim is what separates the men from other tall people, probably because modern technology has provided bubbles and balls, and servo indicators and wheel alignment scales and all manner of gimcracks to confuse the basic issue. Essentially your aircraft is in trim when the sum of all yaw forces is zero. When that happy balance is achieved, the airplane will fly straight ahead (with the wings level) without any long-term heading deviations. It doesn't matter two figs what all the bloody gimmicks tell you as long as the heading remains constant with the wings level.

Roll trim is pretty simple once the yaw has been tied down. Neutralize the lateral control forces with the wings level and that's it.

There are several ways to proceed, but here's my procedure for cruise trimming, most of which will also apply to any other steady-state flight path such as climb, descent and approach.

221

1. Turn off the yaw damper until you are satisifed with all trimming. I know that series yaw dampers aren't supposed to interfere, but I'm not totally convinced.
2. In multi-engine airplanes, be particularly careful to balance the power/thrust symmetrically, using the most reliable indicators: torque, EPR (Engine Pressure Ratio), N_1.
3. Spend a minute to balance the fuel load across the airplane centerline. This can be done with real precision only if you have a fix on the accuracy of your fuel gauges. Fuel and thrust imbalances can be trimmed out with the flight controls, but the net result will be some added drag.
4. Set the aileron and rudder trim to zero. If the airplane is poorly rigged, you may have to hold considerable wheel and/or pedal forces to maintain level flight. You might even want to check the actual tab positions on the ground to make sure that they are precisely fared with their primary control surfaces when the respective cockpit indicator reads zero.
5. Set the pitch trim to maintain the attitude and/or airspeed desired.
6. Now start with the rudder trim. Roll in enough to center whichever ball or spirit-level bubble you consider most accurate, while using ailerons as necessary to hold the wings level. When you think the trim is close, mark your heading or set the heading bug if you have one. Continue to hold the wings level—by reference to the actual horizon if possible—and check for heading drift over a period of a minute or two. When trimming for a long leg, be particularly fastidious. Just one degree of heading drift per minute is equivalent to a 60-degrees-per-hour imbalance which, if not trimmed out, must be overcome by unnecessary control-surface deflection.
7. When you're entirely happy with the yaw trim, remove any aileron forces with roll trim. Control-wheel position doesn't mean a thing exept that you might want the mechanic to rerig the circuit sometime just for the aesthetics of a centered wheel. Your car's front end can be perfectly aligned with the steering wheel off-center and the same principle applies to airplanes.
8. If you have used some substantial amount of roll trim, check the yaw trim again to make sure that adverse yaw forces from the ailerons have not altered that extra-fine directional balance. When you're satisfied, turn on the yaw damper if appropriate.

FUEL CONSCIOUSNESS

Now the airplane should be trimmed for the existing thrust and fuel symmetry, configuration and airspeed. Check all the indicators for later reference. Whatever they show now is the indication for perfect trim. If they are not zeroed, they need to be adjusted.

During a long cruise segment I like to disengage the autopilot at least once every hour to recheck the trim. During approach I just ignore misrigged throttles and trim out the thrust imbalance so that I can concentrate on more consequential items.

When you finally get that beast in trim, you're gonna love it. Fuel consumption will be reduced some slight amount. Handling will be much improved even for very minor refinements in the total trim blend. When you land in the summer those bug marks on the windshield will be distributed with artistic uniformity across all the windshield. That's not a bad indicator either.

CONSERVATION AND ALTITUDE

Descent profiles can aid fuel conservation measurably. Piston pilots need to monitor engine temperatures carefully, but a good compromise profile will be one of ever-increasing steepness as the power is stepped down to cool the engine gradually.

Turbine-powered aircraft usually will benefit from the steepest possible descent at long-range cruise airspeeds. Such a profile involves a constantly decreasing Mach but a fairly steady IAS (Indicated Airspeed). In most cases, the IAS will fall between 200 and 250 knots.

Below 10,000 feet, fuel conservation is largely a matter of minimizing distance, delaying flap and gear extension, and refraining from in-flight speed-brake use.

World energy supplies are not likely to improve in the foreseeable future. Your best source of additional fuel is going to be solid, professional conservation.

Sometimes proper fuel conservation techniques become a primary safety device. You may have loaded enough fuel for two hours en route, one hour to an alternate and 45 minutes of reserve. Actual circumstances may require you to use some of that 3 + 45 worth of fuel in the terminal area because the gear will not go down. Or you may use some of that 3 + 45 troubleshooting a system immediately after takeoff and then refile to an en-route stop for fuel. Planning is essential but the best-laid plans often go astray, so real-time decisions often take precedence.

FUEL MANAGEMENT

Still, the best in-flight fuel decisions will yield little benefit if the fuel on board is not continuously delivered to the power plants. Witness just these three examples from my own experience.

• Several years ago I returned home from a trip to find a close family friend seated in my living room wrapped in bandages. He had crashed in a single-engine plane just short of the runway, with six people on board, due to fuel mismanagement.

I can testify that this friend is one of the most competent, professional pilots with whom I have had the pleasure of flying. Yet, despite his several thousand hours in single- and multi-engine jets, he simply mishandled that elementary system and starved the engine with fuel still in the wings.

• On the last takeoff of a grueling three-day sequence, the number-one engine of my DC-8 flamed out about ten knots below V_1. The abort was simplified by lots of dry runway and as we taxied clear the flight engineer discovered the problem. In the rush to preflight, fuel and depart, this normally thorough engineer had inadvertently transferred all fuel out of the number-one main tank into its alternate.

• A Cessna Citation had to be repositioned to another airport about ten minutes away. Cockpit fuel gauges indicated a total fuel load of about 1,000 pounds, but both engines flamed out at 1,500 feet, and the pilot landed in a field with minor damage to the aircraft. Contamination of the fuel-tank quantity probes distorted cockpit indications, so that those nearly empty tanks *appeared* to be one-third full.

Fuel management must have a very high priority in the total cockpit regimen. You can mismanage hydraulics, pressurization or electrical systems and still bring the airplane home. But fuel mistakes, even simple ones, are terribly unforgiving. Over the past five years the U.S. Navy has recorded 52 accidents or incidents due to pilot mismanagement of fuel systems. Bear in mind that ten-per-year rate does not include running out of fuel due to being lost or overextended. Those sobering figures include only mishaps that involved a pilot's lack of knowledge about management of the fuel system.

If the Navy can do it, so can we.

You could say that fuel system management is a matter of priorities. Even if it's necessary to divert your attention from other items of flight management, you *must* maintain that vital flow of fuel to the power plant. I like to divide fuel management problems into two broad categories, gauging and plumbing.

Gauging problems are universal. The J-3 Cub, with a float-supported wire that stuck right through the filler cap in front of the windscreen, had probably the most accurate and reliable indicating system ever devised. But even it required careful interpretation.

Newer, more sophisticated aircraft have gauging systems that measure fuel by weight (pounds) as opposed to volume (gallons) in order to minimize the inaccuracies introduced by temperature changes. Some are corrected continuously with automatic temperature compensation.

Still, even the best digital electronic units retain some measurable error. Even the best fuel-quantity measuring systems should be calibrated annually and, in flight, regarded as a broad estimate of fuel remaining. Usually they will be found very accurate. But is "usually" certain enough when so much is at stake?

Failure modes in the gauging system can be insidious, and in this case what you don't know *can* hurt you. Flight manuals do not describe the specific failure problems associated with quantity-measuring systems, so you will have to dig that information out from the manfacturer's technical representative or other informed source.

As a minimum, you will want to know the symptoms of electrical failure in reference to the gauge and probes. In most cases, the most accurate measure of remaining fuel will be your own running tally of fuel consumption based on a visual pre-flight check of tank quantity, experience and simple arithmetic. The clock is still a wonderfully useful alternate fuel gauge.

Failures in the fuel system itself are usually more obvious, if not less troublesome. If your aircraft is equipped with a multiple tank system that requires inflight transfer and/or crossfeeding, you will need to mentally assess the effects of any possible failure. Try to anticipate those failures that will prevent you from using all of the fuel on board.

And when you need it all for long over-water or IFR flight, be sure to check each pump and valve early in the flight before your safety is dependent on their operation.

Any fuel problem is likely to demand your best cruise-control techniques in order to conserve whatever is left. As a minimum, you should know the speeds and power settings for maximum endurance and maximum specific range. In addition, you should know:

- The accuracy tolerance of your fuel gauges.
- The location of alternate landing sites.
- The burnout rate for any normal flight profile.

Fuel is time, particularly when some element of the fuel system malfunctions.

And time is the problem.

A SAMPLE FLIGHT

Good fuel-system management is an operational technique to be developed just like pressurization control or proper engine leaning. Solid habits and some basic knowledge of failure modes will suffice. Fuel planning and in-flight fuel use are more dependent on circumstances than is system management. Consider a hypothetical flight.

The trip is planned from A to B, a two-hour flight. The weather forecast suggests an alternate, C, one hour away from the destination, and the pilot selects one hour of holding fuel in anticipation of terminal delays. He also opts for one full hour of reserve fuel, 15 minutes more than the FAA minimums for IFR, because of weather and other uncertainties. The planned fuel load looks like this:

Burnout	2 hours
Alternate	1 hour
Hold	1 hour
Reserve	1 hour
Total fuel	5 hours

On paper this trip has enough fuel to fly from A to B, hold for one hour, divert to C and land there with an hour of reserve fuel. That amount of fuel seems adequate for a two-hour trip. Actually, this airplane has five hours of fuel on board that can be used in any way that the pilot elects. At engine start-up the fuel clock also is started, and when the last minute ticks off, the propulsion quits.

It's elmentary, to be sure, but some professional pilots seem to have lost that basic perspective.

Our fictional trip departs at 0800. The fuel clock will expire at 1300. By no later than 1200, this flight should be in the landing pattern of some appropriate airport. That last hour of fuel is too precious to be spent on anything but emergencies and contingencies.

At 0940, 1 + 40 after takeoff, the pilot receives holding instructions with an expect further clearance time (EFC) of 1040. He has 3 + 20 left and still has several options, including a return to A with safe reserves. His planning appears to have been solid so far because he can hold for the forecast hour and still have enough fuel to divert safely, but all of the options must be juggled with regard to fuel depletion. For planning purposes the fuel remaining at 0940 breaks down to

Reserve	1 + 40
Alternate	1 + 40
Diversion fuel	2 + 00
Total fuel	3 + 20
Minus diversion fuel	2 + 00
Delay fuel	1 + 20

That is, the pilot now has 1 + 20 before he must head for the alternate. No matter where he is at 1100, he must divert or compromise the safety of flight, presuming that weather conditions are still unfavorable at the destination and favorable at his selected alternate. As those change, so should his plan.

EFCs are, in themselves, one of the more confusing factors in the game of fuel consciousness. They are most valuable as lost communication times, but as planning times they often introduce an element of confusion that can compromise good judgment. The unknown is this: How much fuel-time will be spent on vectors and on the final approach?

We need enough fuel at "the first airport of intended landing" to reach the alternate and still keep the reserves. In our example, we need two hours of fuel when we miss the approach at B if we are to maintain our safety option of a diversion to the selected alternate. That is, we need enough fuel to hold, weave through a maze of ATC vectors and fly the final approach, *plus* two hours. That undefined amount—call it "vector fuel"—often can amount to 30 minutes or more.

At 1030 the flight is assigned a new EFC of 1100, but the controller assures the pilot that this new estimate looks solid. On the surface, this timing may look okay because at 1100 enough fuel will remain to reach the alternate with an hour of reserve. Unfortunately, it doesn't work out that way.

When the flight departs this holding pattern at 1100 it will again begin those delaying vectors to the final approach for B. Each minute, from then to touchdown, will come out of that precious reserve, which ought to be available on arrival at the alternate, C, in the event that a diversion is necessary. Each minute after 1100 spent en route to B is in effect stolen from the one-hour reserve we wanted to maintain at our arrival at C.

In other words, at 1100 there will be two hours to go until the engines stop. Only the pilot can best judge how to use that time, but if the original planning philosophy is still valid, safety will be compromised by an EFC of 1100 if the pilot elects to proceed to his original destination.

When should you declare a critical fuel situation or fuel emergency to ATC? How much fuel-time is enough? Consider the minimum safe arrival fuel for a final approach.

Let's begin with the assumption that every IFR arrival ought to have enough fuel for the instrument approach, a pull-up to a visual circle and a visual approach for the actual landing. The minimum level of safety provides that timeless reserve equivalent to a single missed approach and amounts to about 20 minutes' worth of holding fuel in most aircraft.

That 20 minutes, then, is the minimum you should ever see on the fuel gauges over the outer marker. Accordingly, I recommend that you clearly tell ATC that your situation is "fuel critical" as soon as it is apparent that your approach fuel will be less than 30 minutes and the airport in question is IFR. When it is obvious that your approach fuel will total less that that needed for 20 minutes of holding, you are a flying emergency and should so state your condition to ATC because you are committed to the first landing. With less than 20 minutes of holding fuel, you simply do not have enough reserve for a missed approach.

In light of these figures, the FAA's minimum fuel reserves of 30 minutes' VFR and 45 minutes' IFR appear to be minimums indeed.

It seems fatuous to propose guidelines to avoid running out of gas, but recent accidents suggest otherwise, so here goes:

• Plan carefully and realistically, never proposing less than FAA minimums.

• Know how much flight time your departure fuel will provide and watch the clock as often as you watch the fuel gauges.

• Stay flexible enough to modify that original plan as circumstances dictate. Don't be afraid to return to your departure point or to divert to an alternate if either option is the safest course. Smart pilots do it without hesitation.

228

FUEL CONSCIOUSNESS

• Be aggressive and assertive about declaring a fuel problem or emergency to ATC. That sort of professional determination is important.

• If you have a fuel emergency, take any necessary action to land on an airport. In the end, a controlled landing on an airport takes precedence over every other consideration.

Lindbergh carried fuel reserves for success and fuel reserves for failure. You should do no less.

FUEL QUALITY

Fuel consciousness is an artful blend of planning, utilization and system management, but even those practices are not enough to preclude a fuel problem. You have to have the proper stuff in your tanks to begin with. Fuel quality is the keystone of fuel consciousness.

Ten years ago aviation fuel problems were largely confined to water contamination and proper octane selection. Now—thanks to the forces of geopolitics, more critical engine specifications, shortages and distribution problems—aviation fuel has become a complex subject with increasing implications for the pilot.

One element that bears emphasis is that you, as pilot in command, are responsible for your fuel load. The lineman may pump too much or too little—into any available orifice and out of the nearest truck—but you have to live with it and you carry the legal responsibility for getting the required quantity and quality.

Aviation gasoline (avgas) is a unique product that bears only casual resemblance to automotive gasoline (mogas). Avgas is a precise blend of paraffins, aromatics, naphthemes, olefins, tetra-ethyl lead (TEL) and dyes that conforms to rigid international standards. Its octane rating and lead content are indelibly color coded as follows:

Red 80/87 octane with 0.5 milliliters of lead per gallon.
Blue 100/130 LL (low lead) with two milliliters of lead per gallon.
Green 100/130 with three to four milliliters of lead per gallon.
Purple 115/145 with 4.6 milliliters of lead per gallon.

Octane is nothing more than resistance to detonation and you need at least the minimum octane for the engine in question. Higher octane fuel will always perform well and will not, as a famous myth insists, burn your valves. A higher octane fuel may,

229

however, cause lead fouling of spark plugs. (But that can be minimized by careful leaning.)

Automotive gasoline (mogas) is entirely different. Its octane rating is determined through different testing procedures. Its composition varies widely with locale, and there is no single standard. Despite the experiences of others, I believe it is absolutely unsuited for use in aviation for at least the following reasons:

• Mogas octane ratings will almost certainly be too low for the airplanes used in business aviation.

• Mogas can have a higher vapor pressure because it is not refined for use at altitude. That higher vapor pressure may well cause vapor locks.

• Lead additives in mogas contain extra chlorine and bromine, which are highly corrosive.

You may not like to burn 100-octane low-lead in your 80-octane engine, but it is far superior than any automotive gasoline you can buy.

Jet fuels come from the so-called middle-distillates of the refining process, along with diesel fuel and heating oil. Jet-fuel availability and price have been hit doubly hard in recent years by the fuel shortage and by newer, more stringent specifications that limit the total petroleum fraction available for jet fuel. Newer, relaxed standards are currently under study that would increase the fraction by lifting freeze-point levels. Such higher freeze points have real significance for business and commercial aviation, particularly in light of the increased range and altitude capabilities of newer airplanes.

The freeze point of jet fuel ranges from about −40°C to −60°C depending on its composition. This freeze point has nothing to do with water contamination and is not affected or prevented by fuel heaters or Prist, which are intended to cope only with water crystals (ice). At some critical temperature, jet fuel becomes mushy, like soft wax, and will not flow. If it gets cold enough, it will just sit in the tank like gelatin while the boost pumps cavitate and the engines flame out.

You must know the freeze point of your fuel for any long-duration, high-altitude flight and avoid that tank temperature by descending to a warmer altitude if necessary. Jet-fuel freeze could become a serious threat in the future.

Two common problems shared by avgas and jet-fuel users are contamination and mixing. Contamination is invariably a water problem and easily avoided by careful sump draining. Remember, though, that it is best to drain fuel sumps *before* fuel is added, rather than immediately after, because the mixing action will stir up any water on the tank bottom and leave it in suspension for a considerable time.

Mixing avgas and jet fuel is a more insidious problem that poses at least two serious possibilities.

First, mixing jet fuel into avgas is a recurring problem in piston-powered aircraft that seems to be aggravated by "turbocharged" decals. Some line personnel just don't know the difference between turbocharged and turboprop. At least one instance of putting jet fuel into a turbo Navajo has occurred. The Navajo is an airplane that happens to closely resemble its turboprop brother, the Cheyenne.

Any blend of jet fuel and avgas will have a dangerously low octane rating. Depending on the ratios involved, damage will range from minor to catastrophic. Never fly with *any* amount of jet fuel in your avgas.

Second, there's a danger from putting avgas into jet fuel in a turbine-powered aircraft. Virtually all turbine engines will run all right on avgas or jet/avgas mixtures, but there is a dangerous side effect. The vapors from this combination can be exceptionally explosive. Kerosene has a high flash point, which makes the vapors in a jet fuel tank too lean to explode. Avgas has a low flash point so that its vapors are too rich to explode. When you mix the two, there is a crossover point at which the air-fuel mixture is optimum for burning and any suggestion of a spark will trigger an explosion. The flight manual may allow you to burn some amount of avgas in your turbine engine, but be aware that the mixture can be dangerous in the tank.

It would be nice if we could assume that aviation fuel is an innocuous, helpful fluid. Unfortunately, that assumption can be dangerous. And the prospect is that things will get worse before they get better.

Altitude Awareness

39

One item, still available in bountiful supply, is altitude. You can have as much "up" as you want. It's free and it's available but some pilots are not taking advantage of it. It is not a new problem and it is often misunderstood in and out of the aviation community. During the 1930s, the U.S. Army Air Corps grew from a small extension of the Signal Corps to a significant military force. That rapid expansion in size and technology created unique problems and the accident rate soared. General Henry H. ("Hap") Arnold, concerned by the high rate of aircraft accidents, appointed a committee to study the problem. Unfortunately, his choice for chairman was a cavalry officer; after careful research and study the committee reported that "the primary cause of aviation accidents was aircraft striking the ground."

Antoine Marie Reger De Saint-Exupéry is a popular folk hero in France. In the late 1920s, Saint X (as he was called) was a pilot with the early air-mail service in Argentina, flying the mail through the Andes mountains in all weather, even at night. He developed a firsthand appreciation for altitude on those lonely flights over irregular and deserted terrain, and he wrote of those experiences in his classic book *Night Flight*.*

Now the Patagonia mail [plane] was entering the storm and Fabien [the pilot] abandoned all idea of circumventing it; it was too widespread for that, he reckoned, for the vista of lightning flashes led far inland, exposing battlement on battlement of clouds. He decided to try passing below it, ready to beat a retreat if things took a bad turn.

---

*Excerpted by permission of Harcourt Brace Jovanovich, Publishers, New York, © 1974.

He read his altitude, five thousand five hundred feet, and pressed the controls with his palm to bring it down. The engine started thudding violently, setting all the plane aquiver. Fabien corrected the gliding angle approximately, verifying on the map the height of the hills, some sixteen hundred feet. To keep a safe margin he determined to fly at a trifle above two thousand, staking his altitude as a gambler risks his fortune.

Some basics never change. Aviators are still staking their altitude as a gambler risks his fortune. Sadly, so many are losing that safety experts and accident investigators had to coin a new term: controlled flight into terrain, or CFIT. The accidents are not new, however. The definition of CFIT is as bureaucratic as its nomenclature: "an aircraft in normal flight regime, with no emergencies and no warning to the crew of impending trouble, impacting the terrain at some place other than the runway." The government's way of saying, "flying it into the ground."

Not all lapses in altitude awareness result in CFIT, but many do and others come perilously close. Consider the following:

• In December 1962, an Eastern Airlines Lockhead L-1011 arrived at Miami with an indication of a minor landing-gear problem. While holding at 2,000 feet above the nearly sea-level terrain, the cockpit crew of four concentrated on resolving the landing-gear problem. In fact, they concentrated so much on that minor glitch that no one noticed the airplane's gradual descent until the 1011 mushed into the Everglades, west of the airport.

• On December 1, 1974, TWA 514, a B-727 inbound to Washington's Dulles International Airport struck a mountain after prematurely descending below the transition altitude from the initial approach fix. Ambiguities in the pilot-controller communication terminology and in the applicable charts contributed to this tragedy, but the cockpit crew had received alarming indications of minimal terrain clearance. Bickering between the Airline Pilots Association and the Professional Air Traffic Controllers Association largely obscured a basic lack of altitude awareness.

• In 1973, Delta 723, a DC-9, crashed just short of the runway threshold in Boston. The NTSB summary read, in part: "The probable cause of the accident was failure of the flight crew to monitor altitude and to recognize passage of the aircraft through the approach decision height during an unstabilized approach."

These and dozens of other CFIT accidents were caused by compound "system" problems. And yet, each one could have been prevented by pilot vigilance.

ALTITUDE AWARENESS

Now I know that this is controversial ground. The concept of the pilot as the ultimate accident preventer has fallen into disfavor in our increasingly permissive society where affixing personal blame is unpopular. Still, we pilots are unavoidably positioned between safety and accidents. It is one thing for the Airline Pilots Association to reject the notion that pilots can cause accidents. It is quite another to recognize that pilots can indeed *prevent* accidents.

When Neil Armstrong made the first moon landing he was forced to assume manual control of the Lunar Lander in the final seconds of descent due to some serious "system" problem. Although he would not have caused the potential accident he did, in fact, prevent it.

Altitude awareness is an essential element of airmanship. Why, then, do competent, experienced aviators surrender that precious commodity like carefree gamblers? The answers are controversial, subjective and incomplete. In fact, there may not be any real answers, only suggested cautions.

Still, there are reasons for these breakdowns in altitude awareness, and they seem to fall into three major categories: flight management, communications and visual illusions.

FLIGHT MANAGEMENT

How do I define so basic a concept? How do I instruct professional aviators on their responsibility for basic leadership, basic command?

The fact is, that *I* can't. But *you* can evaluate your own capabilities in this area, and I urge you to do just that.

Pilots in command of any aircraft must assume the total responsibility for planning, directing, and supervising flight with a view to the final results. Despite recent social trends toward equality in all things, there is no room for democracy in the cockpit. Someone must be willing to exercise authority, and when the command pilot demurs, that vacuum will be filled by dispatchers, subordinate crewmembers or ATC, often to the detriment of flight safety.

Many accident reports indicate that command pilots are willing to exercise their authority but are unsure of their proper role in flight management. As the accidents cited earlier demonstrate, one frequent scenario finds the entire crew so engrossed in a routine systems problem that basic airmanship is virtually ignored. Such incidents point out the need for an essential element of management: delegation, the act of entrusting subsidiary responsibility to an assistant so that the manager is free to pursue major objectives.

Flight management often breaks down at that point of delegation, apparently because pilots are trained to believe that leadership demands personal involvement. The truth is that since command pilots are responsible for the overall results of flight their limited capacity for attention must be rationed carefully. And that precious personal attention must be allotted first to the priorities of altitude, airspeed, fuel and navigation. Pilots in command must learn to entrust lesser responsibilities to competent co-pilots and flight engineers, especially when total workload begins to encroach on those four critical items.

In all cases the cockpit must be managed by an alert and assertive leader, because such management is essential to effective altitude awareness. Procedures and checklists—and even emergencies—must not be allowed to obscure the fact that someone must always mind the store. An aircraft without persistent leadership is unsafe at any altitude.

COMMUNICATIONS

Effective communications between captain and crew are essential to altitude awareness for two reasons. First, solid communication draws each cockpit crewmember into a running mental awareness of vertical position. Second, it is the means by which altitude awareness is reinforced.

Making a mental note of altitude information and verbally sharing it with other crewmembers is less important during cruise, more so during climb and descents and absolutely vital during approach.

In VFR cruise you need only ascertain from your chart the minimum en-route altitude (MEA) and avoid descending below it. Most often, a simpler precaution will suffice. At 16,500 feet you will clear all terrain in the continental United States by 2,000 feet; 8,500 feet will clear all terrain east of the Rocky Mountains by an acceptable margin. If your flight planning time does not allow close attention to MEAs, those two simple altitudes can keep you away from ground obstacles. Naturally, cruise altitudes must conform to ATC assignments or the VFR odd/even rules, but most pilots do not have serious trouble with that.

During the initial climbout, altitude awareness must be tempered with a working knowledge of the local terrain. That topographical information can come from area charts, personal knowledge, or the Minimum Sector Altitudes (MSA) circle on airport approach charts. MSAs are relatively new and there is reason to believe that many pilots are not routinely taking advantage of these very useful data.

MSAs are published in the upper right-hand corner of most Jeppesen approach charts, and on the plan view of the Government's NOS charts. MSAs, however, are not designed for all approaches. The altitudes shown in the MSA circle provide at least 1,000 feet of clearance above the highest obstacle in that sector within a 25 nm radius of the reference facility. As many as four sectors, each with its own altitude, may be depicted, or the circle may be ascribed a single altitude.

MSAs are a convenient guide for arriving and departing pilots. They provide basic terrain clearance information that can be incorporated easily into appropriate checklists and crew briefings, underlining the need for solid cockpit communication. Remember though, these altitudes do not necessarily assure acceptable navigational signal coverage.

Other sources of terrain information for departure are as close as the en-route chart and the tower or departure controllers. In all cases, flight crews should include safe terrain-avoidance altitudes and routes in their predeparture briefing, particularly at unfamiliar airports.

As altitude decreases during the descent and approach, communication assumes even more importance. Accordingly, crew communications should increasingly focus on that critical element as the aircraft nears ground level.

Descents from jet flight levels can be particularly critical because the extended time involved can induce complacency, although the problem is not exclusive to high-altitude flying. Turbocharged piston-powered aircraft are best descended at painfully slow rates to avoid major power reductions that can fatigue exhaust components and turbine wheels. Those long, slow descents require a constant awareness of vertical position.

In all cases, crewmembers should verbally note 10,000-foot increments during descent. At FL 180, altimeters are set to the local altimeter setting, so that altitude also becomes a benchmark on the way down and should be so noted by a verbal exchange.

Ten thousand feet msl is a critical point during descents performed at pilot discretion from above 20,000 feet because 10,000-foot increments are easily confused. That last 10,000 feet is utterly unforgiving when mistaken for some higher increment, particularly at the high rates of descent common to turbine-powered airplanes.

When the destination airport's elevation is significantly higher than sea level, you must transition your thinking to absolute altitude—terrain clearance—at some point during the descent. That mental change is best accomplished at 5,000 feet above airport elevation, and crewmembers should verbally note that vertical point.

As the approach begins, altitude call-outs become increasingly important beginning with the initial approach altitude or pattern altitude, as appropriate. One thousand feet above airport elevation is another key gateway, as is the final 500 feet.

During approaches in IMC (Instrument Meteorological Conditions) it is also important to announce and confirm 100 feet above minimums. The last mandatory altitude communication should be a clear confirmation of DH (Decision Height) or MDA (Minimum Descent Altitude).

All of these verbal exchanges may appear to clutter the cockpit workload, but experience proves otherwise. Several major airlines require such exchanges and have achieved excellent success with them. In fact, all of the recommended altitude communications require mere seconds and simple one- or two-word responses. Consider the sequence for a flight descending from 35,000 feet to sea level:

At FL 350—"Leaving 35,000."
At FL 300—"Leaving 30,000."
At FL 200—"Leaving 20,000."
At 18,000 feet msl—"Altimeter 30.02."
At 10,000 feet msl—"Leaving 10,000."
At 5,000 feet agl—"Leaving 5,000 feet above field altitude."
At initial approach altitude or pattern altitude—Announce actual altitude above ground.
At 1,000 feet agl—"Leaving 1,000."
At 500 feet agl—"Leaving 500."
At 100 feet above DH or MDA—"100 feet above minimums."
At DH or MDA—"Minimums."

All of the recommended altitude call-outs total about 40 spoken words. These 40 words will require a mere 15 seconds out of a period of at least 20 minutes, surely a worthwhile investment of precious time.

ALTITUDE AWARENESS

Single-pilot aircraft must do without the luxury of verbal exchange and reinforcement, but the principles of altitude awareness remain unchanged and the call-outs should be made. Pilots flying without a co-pilot should verbalize those vertical checkpoints to maintain their own awareness.

Regardless of the exact wording, crewmembers should freely exchange altitude information on a prescribed and regular basis. Command pilots must insist on strict adherence to those procedures. Altitude information is far too important to be kept secret.

VISUAL ILLUSIONS

Every flight terminates in visual conditions, so the final and most critical altitude judgments are made largely by sight. But visual and perceptual illusions can subtly deceive even the most experienced pilot. The fact that these illusory effects are a significant cause of approach and landing accidents, even in sophisticated aircraft, highlights the universal nature of the problem. In unusual situations human visual perception in particularly subject to gross error. You remember from Chapter 7 but be sure to appreciate how easily those illusions can compromise your altitude minimums.

THE AUTOPILOT APPROACH

Sometimes the best defense against visual illusion is the auto-coupled approach, using the autopilot coupling modes to fly the airplane down to minimums.

Autocoupled approaches markedly change the pattern of cockpit business and introduce several unique and critical failure modes. Autopilot operation is as much an art as is good stick and rudder work, and you need to begin with an intimate knowledge of the equipment in your airplane. Sometimes that information can be difficult to obtain because autocoupler design is still largely a black art. One designer we spoke to freely admitted that he didn't believe in telling pilots how the thing worked because his job was to exclude pilots from the loop anyway. Be persistent, get the information somehow, because you must understand the hardware in order to capitalize on its idiosyncrasies. The best advice you'll ever get on how to fly with autopilots is: do not let yourself become a passive button-pusher.

239

COMMONALITY

There are many approaches to autocoupler design, but there are some common elements. If you are flying a high-performance airplane with a sophisticated coupler, it works something like this: Up to the point of localizer and glideslope intercept, the autopilot functions as a stability platform to keep the aircraft in level flight at some preselected altitude. Localizer intercept is usually possible up to 45 degrees, and at least 20 degrees of intercept angle is desirable to ascertain that the beam has been captured. With some autopilots, glideslope capture is possible only from below. Know those characteristics of the one, or ones, you fly. Be especially cautious of coupled approaches initiated from an on-course, on-glidepath position because you will have no intercept maneuvering to confirm that beam capture has triggered the appropriate autopilot responses. I have seen couplers engaged from a needles-centered situation only to find out too late that one channel or the other was not really engaged. The airplane was tracking near center line only by coincidence.

Localizer capture is pretty straightforward, although there may be some scalloping as the electrons adjust to crosswinds, airplane trim, and signal idiosyncrasy. Glideslope intercept is often a two-stage process involving an initial pitch-over to some specific rate of descent followed by actual beam capture. Autopilots that use that logic need to be carefully monitored to assure that sequencing doesn't fail after the initial push-over. If conditions are just right, that primal rate of descent can keep the airplane very close to the glidepath until it's too late to reengage. In both cases, allow the autopilot some pretty substantial oscillations during the first 30 seconds after intercept for normal adjustments to wind, temperature, airplane trim and speed. During that half minute, look for progressive stabilization in both axes and recheck the annunciation panel for appropriate indications.

After capture, continue to keep a skeptical eye on mode indications. Some autopilots will silently uncouple if the navigational signals are momentarily interrupted or if displacement from the on-course position exceeds certain limits. We have experienced that sort of uncoupling due to a large truck passing near the ILS shacks and distorting the signals.

Somewhere around the outer marker your equipment may begin a timed gain reduction of the LOC and/or G/P (Glide Path) signals. Autopilot engineers call this feature attenuation, and it is used to reduce signal sensitivity so that coupled responses are progressively softened as those beams narrow down to critical size. Attenuation is good for you, but it's even better if you know where it begins and what happens if it doesn't. Depending on the manufacturer and model, attenuation may begin at glideslope intercept, middle-marker reception, some predetermined value of radio altitude, or not at all. Each of these different configurations presents its own failure possibilities.

GLIDESLOPE EXTENSION

Glideslope programming may include another little goody, known as glidepath extension. By the time you get to 200 feet or so, the glideslope beam is so narrow that any aberration in the signal could cause a gross disturbance to the airplane's flight path. Glidepath extension is simply a programming device within the autopilot computer that allows a memory circuit to complete the last few hundred feet of vertical guidance.

So what can you do about it after you have researched your own equipment? You can practice and learn how to tweak those black boxes for optimum performance. We have found several tricks that allow the pilot to maximize the airframe/autocoupler combination.

It is always necessary for the autocoupling procedure to initiate well within the prescribed speed envelope because some autocouplers just don't have the authority to correct for gross speed changes and none of them do it well. Try to give the autopilot its best shot at flying a smooth, stabilized approach. Avoid large flap movements for the same reason.

Once established in the approach, you may find that the coupler functions appreciably better at a speed ten to 12 knots above minimum weight V_{REF} and that the added speed is justified for this automated operation. For instance, a Sperry SP-30 instrument does its best work at or above 140 knots, although V_{REF} can be as low as 120 knots.

During the approach you have no direct control over localizer or glideslope tracking, but it is possible to cheat a little when the coupler needs persuasion. Asymmetrical thrust, or just a foot on a rudder pedal, used judiciously, will often coax the airplane over to an on-track condition when the coupler insists on tracking to one side or the other. Similarly, small glidepath displacements can sometimes be corrected by increased or decreased power, especially in those airplanes with a substantial pitch response to power changes. Both of those deceptions are increasingly effective as attenuation progresses.

As you approach DH, mentally establish the runway picture you can expect, based on crab angle and on-course location. When the runway is in sight, resist the urge to disconnect until you have taken a moment to become familiar with the visual environment. You are your own worst enemy for the first few seconds.

Below DH, the pilot not flying ought to announce speed and sink rate until the wheels are on. After touchdown, he should be ready to track the localizer during rollout in the event that fog, rain or blowing snow obscures the runway.

Don't ignore that autocoupler. Start practicing now, and by the time you really need it you will probably understand each other.

DESCENT

Actually, the final approach is merely the conclusion of a much longer descent from cruising altitude. Altitude awareness is perhaps most important during this final stage of flight but descent profiles are often poorly thought out. Navy pilots of my generation will remember a monument to poor descents at North Whiting Field. It was called "Kiwi."

That wrinkled T-28 sat on the edge of the field across from the parachute loft with its landing gear anchored in concrete where it was used as a procedures and bail-out trainer. The story of Kiwi was always shrouded in folklore, but the most commonly told story went like this:

Navy Cadet Jones launched out on a dual instrument-training flight, flying the big Trojan from under the backseat hood. In the process of a jet penetration maneuver from 18,000 feet, his feeble instrument scan overlooked the airspeed until that instrument's needle twirled around to the never-never position.

When the deafening slipstream noise finally captured his attention, he attempted to rectify the speed excursion with an immediate and total speed-brake extension. Although his basic thinking was sound, the pitch-up forces from that underside speed-brake pitched Navcad Jones into a dozen-G pullout which permanently warped and grounded the unfortunate airframe named Kiwi. No one knows what happened to Jones.

Descent from altitude, in any airplane, is a challenging task that has never received the attention it deserves in flight manuals or training programs. At least four factors should directly influence your descent planning.

Efficiency. Jets and turboprops can benefit most from well-planned and well-executed descents, but even piston-powered models can achieve some increased fuel efficiency with proper technique.

Basically, turbine-powered airplanes will consume minimum fuel when descended at long-range cruise airspeed/Mach with the power pulled back to idle.

In the operational environment, that means you must know your airplane's idle-descent capability and you must know it in terms of nautical miles traveled per 1,000 feet of altitude loss. Look at it this way.

As you approach your destination, there is some magic point from which it is possible to close the throttles and glide down, at maximum efficiency, to the first crossing restriction. If you know the nautical-miles-per-1,000-feet rate for your airplane, everything else falls in place easily. My airplane falls at an average round figure of 2.5 miles per 1,000. Ten thousand feet in 25 miles. Twenty thousand in 50 miles, and add eight to ten miles for decelerating from red-line to 250 knots at 10,000 feet.

If you don't know your airplane's descent numbers, find them out on the next descent with accurate distance information. My guess is that three nm per 1,000 feet is a good starting point, but remember that maximum efficiency is achieved at long-range cruise speeds in a perfectly clean airplane. Speed brakes, gear, flaps and high Mach numbers may steepen the descent, but they all increase burnout.

Piston engines should be leaned during descent for efficiency, but also for engine protection and that is an entire subject unto itself.

Engine protection. After that gentle, gradual, total power reduction at the point of descent, there is nothing to be done for engine protection in a turbine. Pistons are different.

The key to piston-engine protection is gradual temperature changes. That means slow power reductions, gradual descents and careful leaning to maintain enough heat in the cylinders. Normally aspirated engines should be operated at minimum cruise rpm and at least 15 inches manifold pressure. Lean the mixture for smoothest possible operation to minimize plug fouling and maintain cylinder heat. Turbocharged engines demand much more care and planning.

243

One major engine manufacturer recommends the following procedure for descent in turbocharged airplanes. At normal cruise power, push over into a 1,000 fpm descent while monitoring the cylinder heads for a 40°C drop. This first, and most critical, 40° reduction should take at least two minutes, so you may have to adjust the vertical speed accordingly.

When the heads have cooled at least 40°C, make normal and gradual power reductions, but always maintain minimum cruise rpm and at least 25 inches of manifold pressure. Lean for smooth operation throughout the descent.

Gradual cooling is beneficial in all piston engines. It may seem tedious in flight, but the alternative is scored cylinder walls, spark plug fouling, cylinder head cracks, warped valves and valve-port damage.

Passenger comfort. Thirty years ago the airlines used a standard descent rate of 300 fpm for passenger comfort in DC-3s and that is still an excellent target for cabin-descent rate.

To find how far out you must begin the descent, multiply your altitude above airport level, expressed as a flight level (7.0 for 7,000 feet, for instance) times groundspeed and divide by 18. At 7,000 feet and 180 knots, for example, $7 \times 180 \div 18 = 70$ nm. At 9,000 feet and 150 knots the distance is 75 nm.

Pressurized airplanes should also aim for 300 fpm cabin rate and that may be the limiting factor for some descents. Figure the difference between cruising cabin altitude and destination elevation. Divide that difference by 300 fpm to find the minutes you must allow for descent. Then minutes times groundspeed divided by 60 will yield the distance needed for descent. Or you can just as easily use the formula above, cabin altitude rather than actual.

Operational considerations. These may alter your normal descent profile. Your airplane will take more horizontal distance to descend when it is heavier. Anti-ice or air-conditioning and pressurization needs may preclude the use of idle power, further extending the glide. Finally, rough air may require an early slowdown with some change to the average profile.

A pilot who should have known better once told me that he didn't have any trouble with descents because after he pulled off the power, gravity did the rest. It's a clever comment, but there is more to professional flying than that.

Altitude awareness. This is essential to flight safety because an unexpected impact with the ground is usually disastrous. Altitude awareness is also closely related to fuel consciousness, particularly in turbine-powered airplanes.

In the preceding chapter, we discussed the ramifications of fuel reserves. In high-flying airplanes, altitude often becomes a significant variable in that fuel equation.

At any given altitude an airplane has two basic energy reserves: the fuel available to power the engines *and* the potential energy of altitude, which can be converted to kinetic energy in a descent. Many business and commercial jets are capable of gliding at a ratio of 15:1 or better. At 41,000 feet, for example, the potential energy of altitude alone will carry that airplane well over 100 nm.

It is important not to surrender altitude casually. Not only is altitude crucial for terrain clearance, but it also can be a significant element in fuel management, particularly in the arrival phase, where delays at low, fuel-inefficient altitudes can be costly.

Altitude is a strange commodity. It is the very thing that distinguishes flying from all other modes of travel. It is the basic margin of success for all kinds of flight. It is the line between safety and disaster.

Still, pilots are freely relinquishing that precious commodity to the pressures of expediency, complacency, traffic control and convenience. Saint X would say they are gambling. Statistics say that the odds are not good.

Cabin Pressure

40

Aircraft cabin pressurization is the system which allows you to carry yourself and your passengers to higher altitudes. Pressurized airplanes may be less prone to CFIT accidents, but they introduce other, definite challenges.

Twenty years ago you could count the number of general aviation pressurized airplane models on one hand. The latest count shows almost 50 pressurized models, from high-performance singles to wide-body executive jets. This decided trend toward pressurization at all levels of general aviation is based on several reasons. Cabin pressure provides a virtual catalogue of benefits:

• Passenger and crew comfort is enhanced because programmed rates of climb and descent almost eliminate any discomfort caused by pressure changes. Also, the altitude flexibility afforded by pressurization increases your chances of finding clear, smooth air.

• Weather and terrain clearance can be accomplished without the need for lengthy and costly detours.

• Fuel conservation is improved by the availability of higher, more efficient altitudes.

• Cabin noise is greatly reduced by the added fuselage sealing necessary for pressurization and by the positive pressure in the hull, which acts as a barrier to external noise.

• Passenger health problems—such as emphysema, heart damage, or sinus infection, which might prevent travel by air—can be neutralized.

Cabin pressurization is a tremendous advantage in any airplane, but it does present the pilot with one more possible system failure with which to cope. A loss of pressurization can range from a nuisance to a catastrophe, depending on the type of failure and the circumstances. In any case, if you are flying—or contemplating flying—a pressurized airplane, consider the possible effects of a pressurization failure. Knowledge and preparation are the best insurance.

The theory of cabin pressurization is simple. Consider the hypothetical fuselage, sealed against all air leaks at sea level. As the passengers embark, the ambient pressure inside and out is 14.7 psi. When the door is closed, the inside pressure in our perfectly secured airplane is still 14.7 psi, the same as it is outside, so there is no differential pressure. If we raise this sealed fuselage to 10,000 feet, the pressure outside falls to 10.0 psi, but because there are no leaks in this airplane, the inside presure remains at 14.7 psi, a 4.6-psi differential.

Obviously, our hypothetical fuselage would have some serious drawbacks, even if it were possible to build such an airtight cabin. For example, the occupants of this sealed tube would be quite uncomfortable because there would be no ventilation. In practice, wings and engines, pushrods, wires and cables produce unavoidable leaks where they penetrate the pressure hull, so that some provision must be made for that predictable leakage. And life requires a constant supply of fresh air.

Airplane pressurization systems provide ventilation and replace leaked pressure by using a constant air input either from so-called cabin superchargers or from bleed air taken directly from turbine engines. This air input is regulated at some constant rate and measured in pounds per minute. Regardless of altitude, airspeed, cabin altitude or engine-power setting, the mass airflow into a pressurized cabin is constant.

Pressure in that cabin is controlled by an outflow valve. If the valve is wide open, all of the mass airflow blows right back out to the atmosphere, and there is no pressure differential in the cabin, just a great volume of ventilation. If the outflow valve were jammed shut, the constant flow of air into the cabin would steadily increase pressure until it reached unacceptable levels. It's like blowing air into a paper bag. If it is torn, all of the input rushes directly out through the hole. If the bag is without holes, pressure increases as the air is forced inside.

In practice the outflow valve is controlled automatically to maintain some selected pressure in the hull, rather like a *controlled* leak in the paper bag. When this control system adjusts the valve so that more air is entering the cabin than is leaving, cabin pressure increases (cabin altitude decreases). When the control system allows more air to exit through the outflow valve than is entering the cabin, pressure decreases and cabin altitude increases.

As you can see, cabin pressurization is really a simple concept with four basic components:

- The aircraft fuselage itself (the paper bag).
- Superchargers or engine-bleed valves for a source of pressurizing air (your lungs).
- An outflow valve to regulate the volume of existing air (a controlled leak).
- A pressure controller to regulate automatically the outflow valve for smooth, coordinated cabin altitude adjustments (the system's brain).

Pressurization systems have proven to be reliable and durable systems that experience few serious failures. Still, problems can occur in any one of the four elements. When that happens, the situation can be seriously complicated by operational circumstances.

AIR SOURCE FAILURE

When the source of pressurizing air fails, the cabin rapidly leaks to ambient atmospheric pressure. Most pressurized twins have a source from each engine so failure of one source merely reduces total airflow with a proportionate reduction in ventilation. Simple failure of one supercharger or engine bleed is not serious. You may wish to curtail passenger smoking because the cabin will not ventilate well at altitude. You may also consider descending to an altitude where the effects of losing air would be lessened.

If your aircraft is pressurized from a single source, failure of that air supply can be critical and may require an emergency descent. We'll talk about that later.

Sometimes superchargers fail in a manner that fills the entire pressurization system and cabin with smoke. This can be the most serious emergency involving an air-source failure.

Three years ago, a Beechcraft King Air A90 lifted off from Cleveland's Hopkins Airport on a deadhead trip with one pilot aboard. At 3,000 feet the left supercharger seal failed, allowing engine oil to contaminate the system and filling the airplane—in seconds—with dense smoke.

The experienced pilot's initial reaction was, quite understandably, confusion. Was the smoke electrical, hydraulic, engine or interior? There was simply no easy way to find out so, as a precaution, he turned off all electrical power before he returned to Hopkins and landed safely. On final approach the pilot was beginning to succumb to smoke inhalation, and the landing was made at about his limit of endurance although total elapsed time was only a few minutes.

An air source that smokes the cabin can seriously jeopardize flight safety. When that happens, follow the handbook procedure. But as a minimum:

• Get on oxygen and use smoke goggles. Any smoke is toxic enough to cause physiological problems.

• Declare an emergency so that ATC will know your problem and monitor your actions.

• Shut off the air supply and manually open the outflow valve to ventilate the cabin.

By this time you should be rid of the immediate smoke problem. Circumstances will now dictate whether to descend and land or to troubleshoot the system in hopes of repressurizing. In some cases, alternately reducing engine power to idle may be enough to identify the smoke source.

CONTROLLER PROBLEMS

Pressurization controllers, those little panels in the cockpit, are surprisingly reliable units with rather benign failure modes. Controllers may wander, causing discomfort and some anxiety, but they seldom create an immediate emergency.

In stabilized cruise, complete failure of the control unit will likely go unnoticed because the outflow valve—and the cabin pressure—will tend to remain in place. Your first indication of controller failure is likely to come as you begin to descend.

If you do suspect a problem, check the applicable circuit breakers. Don't forget the circuit breakers for the air data computer if your aircraft has one and if that unit supplies static information to your pressurization controller. Pneumatic controllers may respond to a rabbit punch, but that is the outer limit of inflight troubleshooting on those nonelectric units.

All pressurization controllers require a source of outside static pressure, so that is an unlikely, but possible, source of trouble. If you are in icing conditions and suspect a static source problem, take the applicable anti-ice and deice actions. In any case try the alternate static source.

CABIN PRESSURE

There is not much you can do with a seriously malfunctioning pressurization controller. Circumstances and good judgment are the best guides.

Pressurization outflow valves are really quite simple mechanical devices. Furthermore, you probably will not have any direct cockpit control over this element of the system unless you happen to be flying airline-size equipment. Outflow valves need to be cleaned periodically to remove tobacco tars and residue, which can cause them to stick, but there is no real in-flight fix for malfunctions.

FUSELAGE STRUCTURAL FAILURE

Fuselage structural failure is the worry of every high altitude pilot because it induces rapid or explosive decompression, which demands immediate pilot action to prevent a disaster. Such failures can range from a door seal that deflates to the catastrophic loss of a door or window.

Slow leaks around the door and window seals need not be alarming if the basic structure is sound. Such leaks may be noisy, but they do not require a panic descent or depressurization.

Discernible leaks in the fuselage structure itself should be treated with extreme caution because they may signal an impending failure. Your best reaction is to minimize the stress by reducing the differential pressure. Select an increased cabin altitude and descend to a lower altitude. The rule of thumb is: cabin up, airplane down.

Catastrophic failures at altitude are so frightening and disconcerting that few pilots will be able to cope without some initial confusion. Explosive decompression is everything that the name implies and will create at least:

- A solid fog as water vapor in the cabin air instantly condenses in the rapidly falling temperature.
- A hurricane of wind and debris as the higher pressure cabin air rushes out of the fuselage.
- A deafening sound.

If you regularly fly above 25,000 feet, you should consider seriously some altitude-chamber training to familiarize yourself with the scary sensations of an explosive decompression. You can receive such training at any of several military bases or at the FAA facility of Oklahoma City.

In all cases of sudden decompression it is essential that you don the oxygen mask immediately. All other matters are insignificant

251

in relation to that primary responsibility. Get the mask on. Get the oxygen on. You cannot be of any help to your passengers and crew without an adequate source of oxygen for yourself.

When your mask is in place and delivering oxygen, you must begin an immediate descent to an altitude at which your passengers can survive. Fourteen thousand feet is a good compromise altitude until the emergency has stabilized and you can resume proper coordination with ATC. Fourteen thousand is low enough to protect life and health and yet high enough to clear 99 percent of all the world's terrain.

That so-called emergency descent bears some careful thought. Flight manuals invariably prescribe a procedure that involves maximum airspeeds and rates of descent using all available deceleration devices. The textbook procedures are written to satisfy the applicable FARs, but they are not always the safest action. Maximum emergency descent is a reasonable maneuver so long as the basic structure is sound, because it expedites the return to a habitable environment. If any part of the structure is damaged or weakened, however, that maximum descent could precipitate a disaster.

Any structural difficulty that affects cabin pressurization should be carefully assessed before committing to a steep, high-speed dive. If in doubt, execute the descent at reduced air-speeds and accept the increased time. Only those few aircraft that fly above 41,000 feet need to expedite the descent at all costs. In reality, this particular problem lends itself to careful thought and rational common sense. Don't be trapped by a rigid procedure. Fly the airplane to a lower, safer altitude by the most appropriate means for the circumstances. If the airframe is damaged, be very gentle. If structural integrity is not impaired, pull out all the stops.

Actually, circumstances are the most important elements in any pressurization problem. Failures on the ground are utterly benign. Failures at 51,000 feet may be fatal. Circumstances rule.

One special circumstance that deeply compounds the pressurization failure is a long, over-water flight. When you flight plan a long, over-water leg at max altitude because you need the best range, you may be betting the airplane on the pressurization system. Look at the numbers.

Consider a typical heavy business jet traveling from California to Honolulu against average winter winds of 90 knots. We'll assume a crew of two, plus eight passengers in back. Cruise is planned at maximum altitude for best range.

CABIN PRESSURE

The climb to cruise altitude is entirely normal and the flight proceeds to the point of equal time (PET) without difficulty. At the PET, the pressurization fails, and a descent from planned cruise altitude is required. Now our trip is in serious jeopardy. The descent for environmental reasons will reduce the airplane's range below that which is necessary to reach Honolulu. Even the onboard oxygen will not be enough.

Our typical airplane is equipped with a factory-standard oxygen installation of two 48-cubic-foot bottles. The total oxygen required for a reasonable, safe flight profile following the pressurization failure is 227 cubic feet, considerably more than twice the available supply. This airplane would be forced to ditch because of a pressurization failure.

When you do flight plan a maximum-range, over-water trip, there are three ways to cover the worst case of depressurization:

• Take enough additional oxygen to supply all occupants at 25,000 feet from PET to destination. Some flight departments have been doing this for years.

• Carry enough fuel to complete the trip from PET at 12,000 feet. Normally this is not an option in business aircraft.

• Restrict the passenger load so that installed oxygen will suffice. In our example, the pilot would have had to limit the trip to three passengers.

Cabin pressurization failures are seldom disastrous. A review of the NTSB records reveals fewer than ten serious incidents or accidents attributed to this cause since 1962. But failures can happen and can compromise the safety of your flight if they are not handled properly.

Sooner or later your pressurization system will act up. The chances of that failure's being serious are sliver-thin, but prior knowledge of the pressurization system and its failure modes will avert serious discomfort and difficulty.

Lessons and Laughs

41

Flying is a serious business, particularly during the en route segment. It's hard to get hurt at the flight-planning desk but in flight you will always need enough runway, enough fuel and enough altitude.

Still, there is some wonderful humor in aviation and some of that humor tells a useful message.

The headline read: "FAA Investigating Midair Collision Between Surfboard and Airplane."

Witnesses reported that a home-built sports plane was buzzing within a few feet of the water off the northern coast of Oahu, in the vicinity of several surfers. One of those surfers saw that aircraft heading toward him and tumbled off the rear of his board. When he did so, the surfboard shot out of the water and struck the biplane's lower wing.

Immediately after the incident the FAA received two phone calls. One from the surfer, complaining about the buzz job; one from the pilot, complaining about the person who threw a surfboard at his airplane.

The message to this fearless pilot is that if you want to mess around, expect to get dirty.

You've heard about hot brakes, burning engines and electrical smoke, but how about the Pyrotechnic Pilot? This hero checks in for work with the nicest handlebar mustache this side of Tipperary, complete with a heavy wax job to hold the shape. The stage is set when he slips on his oxygen mask at 100 percent and all that waxed hair spontaneously combusts, as petroleum products are wont to do in the presence of pure oxygen.

Is there a moral to this hairy tale? Could he have anticipated the consequences of oxygen and oil? Is facial hair out for aviators? Or

255

is there some reason for moderation even in the serious pilot's grooming that is virtually dictated by the restrictions of his working environment?

One of my Navy contemporaries experienced a most unusual bird strike—from inside the airplane. After a routine takeoff in his A-4 this young attack pilot felt a stabbing pain in his left leg. The initial jab was rapidly followed by several more. His problem was caused by a small but feisty bird that had stowed away in the rudder-pedal well and come to life at the worst possible time.

Unable to reach into that narrow crevice in the cramped A-4 cockpit, this Navy pilot endured several excruciating punctures in his lower left leg as the persistent featherhead continued its attack throughout the landing circuit. There was no permanent damage, and the brutish bird was dispatched by ground personnel while my friend limped to sick bay for a tetanus booster.

Many bizarre events have resulted from a glitch in electrical wiring, but would you believe that a single instrument-light rheostat could undo both engines on a turbine twin? It happened to an Air Force helicopter immediately after a minor electrical modification. While cruising at 500 feet the pilot turned up that light rheostat and both engines ran away. A single ground wire connected to the wrong terminal resulted in both fuel controls going to the manual position when the rheostat was turned up. The two pilots regained control of their aircraft and governed fuel manually for a safe landing.

The moral here has to do with that classic, Murphy's Law. There's not much you can do about it except to fight like crazy against complacency.

How about a near midair collision with a pig? A 50-foot-long pig at 1,000 feet? It happened, over London.

Visibility was a mere two miles as the helicopter chugged up the Thames at 1,000 feet. Suddenly, and without any warning, a 50-foot pink porker rose out of the haze layer at one o'clock, less than a mile away. The pilot took evasive action and later filed what the British call an air-miss.

Safety investigators found that the sailing swine was really a publicity balloon that had broken its mooring.

And speaking of animals, the catalogue of bizarre aerial incidents is replete with wildlife:

• A Boeing 737 that collided with a cow just *after* takeoff. My wife has been saying all along that beef is too high.

• We've already talked of bird strikes. One pilot reported a snake strike at 800 feet agl during the climbout. No one is really sure how it got to 800 feet but some have speculated that it must have been dropped by a hawk.

• Rabbits are a nuisance at many airports but the Flight Safety Foundation reports a case of a collapsed gear due to hare avoidance. While taxiing at 40 knots, the pilot turned abruptly to avoid a bunny. The pilot's sentimentality was understandable but his priorities were lousy.

Sometimes those absurd hazards come *from* the airplane rather than happen to it. There have been several instances of so-called blue-ice damage to buildings. The usual complaint involves a large chunk of indigo ice that falls inexplicably through the roof of some buildings.

Blue ice is caused by a slow leak in the exterior toilet-drain fitting on large aircraft. As the colored flush-fluid seeps out, it forms an ice chunk, which breaks away and falls to earth as a blue bomb.

Ironic, isn't it, that a button-down, checklist industry like aviation involves such preposterous antics? Still, my file of operational absurdities grows almost monthly.

Aviation foibles are amusing and often instructive. If there is a common thread to these entertaining misfortunes it has to do with end results. Incidents will happen and the laughs are fun—as long as you're not on the receiving end.

HELICOPTERS VI

The thing is, helicopters are different from planes. An airplane by its nature wants to fly, and if not interfered with too strongly by unusual events or by a deliberately incompetent pilot, it will fly. A helicopter does not want to fly. It is maintained in the air by a variety of forces and controls working in opposition to each other, and if there is any disturbance in this delicate balance the helicopter stops flying, immediately and disastrously. There is no such thing as a gliding helicopter. This is why being a helicopter pilot is so different from being an airplane pilot, and why, in general, airplane pilots are open, clear-eyed, buoyant extroverts, and helicopter pilots are brooders, introspective anticipators of trouble. They know if something has not happened, it is about to.
——Harry Reasoner

Helicopter time in your logbook is like V.D. in your medical record.
——Traditional Fighter Pilot Wisdom

LOW FLIGHT

Oh, I have barely slipped the muddy clutch of Earth
And thrashed the skies on dusty, untracked rotor blades.
Sunward, I've climbed and joined the tumbling mirth of flies and
    bees
And sunsplit crowds of trees—and done a hundred things
    you have not dreamed of—
Wheeled and soared and swung through rocks and bushes
    and wires.
High in the sunlit silence at three feet,
Hov'ring there, I've been chased by crows and sparrows
    and flung my groaning craft along the freeway narrows.
Up, up the long delirious burning blue . . .
I've almost topped 500 feet msl, where never a self-
    respecting lark nor eagle flew.
And, while with silent lifting mind I've trod
The low, smoggy ambience of space
Put out my hand, and searched for FOD.
——Anonymous Helicopter Pilot

Those three perspectives from a layman, an airplane pilot, and a helicopter pilot are representative of the prevailing attitudes toward those very different rotary-wing aircraft.

Helicopters are different. Their dynamic components, flight characteristics, and pilot techniques are truly unusual but the aircraft themselves are increasingly important to the total air transportation system. You may not like helicopters but you would be foolish to ignore them. They provide a unique capability that is often the best possible solution to transportation demands.

Uses 42

Colonial Sand & Stone operates 13 gravel, aggregate-stone and ready-mix concrete plants in and near the New York metropolitan area. To carry those products, CSS operates a fleet of 700 trucks and a navy of 15 tugs and 135 barges and scows from Connecticut to southern New Jersey. But one vehicle, its Bell JetRanger, is the real hingepin of CSS's operation.

CSS's helicopter is strictly a no-nonsense business machine, far more versatile than a pickup truck. Key company personnel use the aircraft to go between outlying plants. Salesmen visit customers over a 100-mile radius. Aerial surveys of new plant sites or portable plants on customers' construction sites save literally days of ground travel and inspection. Trucks and tugs are serviced on location with parts and repairmen delivered by air. CSS's president keeps in touch with satellite plants by twice-weekly visits.

Timex oversees its watch and instrument-making operations from corporate headquarters in Middlebury, Connecticut, 65 ground miles from the nearest airline terminal. Timex's light-turbine helicopter ties that corporation together with a flexible transporation system that avoids the hassle of ground travel. That single aircraft easily earns a profit for the company by moving critical parts and key management people, in that order of priority. In approximately 5,000 flight hours this helicopter has saved nearly 20,000 hours of personal transportation time, while avoiding innumerable production slowdowns or outright stoppages.

Ceasare Bianchi operates a truck repair, towing, and road service operation in Visalia, California. Ceasare's garage lost half of its traffic when Interstate 5 opened in March 1972, and he lost no time in adding a Bell 47G-2 helicopter to his equipment fleet. Now, when a truck is disabled on the new highway, or anywhere else, for that matter, the helicopter delivers parts, tools and mechanics to the site within minutes of the initial call. Because trucking is time-intensive, Ceasare has had no complaints from customers about the additional charges for flying time. This garageman would tell you that his helicopter is the handiest piece of machinery in the operation.

Like it or not, the helicopter is rapidly becoming an essential element in air transportation. If your flight experience is limited to fixed-wing operations, it may be time to take a fresh look at the VTOL potential. When you do take that look, try to keep an open mind because helicopter technology has advanced so rapidly that even those in the industry are having trouble staying informed. Outside that specialized world, misconceptions are rampant. Consider this very incomplete list:

Helicopters are slow. If you think helicopters aren't fast you are probably using an inappropriate speed comparison. Helicopters are really designed to compete against surface transportation. Think of them as 100-knot station wagons operating free from all the constraints of surface routing and congestion. No other vehicle allows the speed and flexibility of helicopter travel. If your company operates a car, truck or boat, it is entirely possible that a helicopter could complement those transportation elements and save money.

Helicopters are expensive. Compared with what? Nothing is expensive if it saves, or makes, money. You can buy six turbine-powered passenger seats in an air-conditioned, fully IFR helicopter for about the price of a pressurized piston-twin airplane. You can order an eight passenger, twin-turbine, 150-knot, fully equipped and instrumented helicopter for about the price of a comparably equipped turboprop. After all, if you can justify a Falcon or a Sabre because they are time-competitive with the airlines, can't you justify a far cheaper machine that is even more time-competitive with a limousine?

Helicopters are uncomfortable. That's another one of those competitive things. Is it more comfortable to ride in a Continental or even a Rolls for two hours than it would be to ride in a helicopter for ten to 30 minutes? If you haven't had a ride in the latest helicopter models, you're in for a treat. During a recent evaluation flight an editor I know ripped off several pages of handwritten copy at the cabin table without a wiggle. Throughout the flight we conversed easily at about the level you would use on the commuter train. And all through the ride we enjoyed 21°C air-conditioned comfort while outside temperatures hovered near 32°C with extreme humidity. How much comfort do you want?

Helicopters are noisy. That statement should read "older helicopter designs were often noisy." The industry has been aware of this problem for several years, and significant improvements are just now entering the marketplace. New rotor airfoils, swept rotor tips, enclosed tail rotors and redesigned turbine exhausts have all contributed to a really substantial reduction in cabin and exterior noise. Even further reductions in cabin noise are possible with custom insulating, and the exterior noise footprint can often be neutralized by simple piloting techniques.

Helicopters are bad neighbors. The truth is that helicopters are good samaritans and better friends when introduced with the proper community-relations attitude. If the town's people are made to understand that your helicopter is an important industrial tool, which is also available for occasional service as a rescue vehicle or ambulance, they may be eager to have it.

But that education process really should start *within* the aviation community because those rotary-winged aircraft have suffered from 40 years of convenient generalizations and blatant misunderstandings on the part of the fixed-wing community.

The truth is, helicopters *are* unique. They are not built, flown, maintained, or managed like airplanes. They are different in design, construction and application, and the astute flight-department manager will want to know how their unusual capabilities can be utilized to his company's advantage.

One significant barrier between the airplane and helicopter communities is terminology. Because they are so different, helicopters and helicopter operations can be described only in terms completely foreign to airplane people. If you expect to shop for a helicopter in the future, you will need to know some basic hardware terminology.

Terminology for the Fixed-Winger

43

BEGIN AT THE TOP

Nothing sets a helicopter apart from airplanes more than its rotor. In fact, the very name helicopter comes from a Greek root meaning spiral or screw, which describes the path of the rotor through the air. "Helicopter" implies a vehicle that literally screws itself into the air, which is not far from the truth.

Rotors have been described as horizontal propellers, but that analogy is very limited. Propellers are designed for the simple task of creating thrust. Rotors, on the other hand, must provide lift *and* thrust while at the same time providing complete lateral and longitudinal control. Rotors are intriguingly sophisticated devices in which the individual blades must be free to move in at least three ways:

Flapping describes the vertical motions of the blades as they swing up and down in response to aerodynamic loads, control inputs, and other forces applied to them during their rotational journey.

Lead and lag or *dragging* or *hunting and dragging* describe the back-and-forth motions of each individual blade within the plane of the rotor disc relative to the steady rotation of the entire rotor. In a four-bladed system, for instance, you might expect that the blades would be oriented 90 degrees from each other. That is essentially, but not always, true. Lead and lag allows some limited individual blade motions relative to the other blades so that each is free to move back and forth out of that precise 90-degree orientation.

Feathering action changes the pitch angle of each individual blade and is the only rotor-blade motion that corresponds to propeller motion. But for a helicopter feathering does not mean the *same* as it does when used in reference to a propeller. For a propeller, feathering refers to the blades being positioned on edge to the relative wind. In a rotor system it refers to changing the blade angle, but not necessarily on "feather edge" to the wind.

ROTOR SYSTEM TYPES

This unusual amount of individual blade freedom demands some extraordinary hardware, and rotor systems fall into four broad categories:

Articulated rotors are just what the term implies. In biology, articulated means jointed and that aptly describes the rotor system; it has many joints. Articulated rotors, by definition, are designed with individual flapping hinges, lead/lag hinges and feathering hinges; a full set for each blade. In other words, each blade is fully jointed about three axes at the point where it attaches to the rotor hub.

Articulated designs are seldom used on two-bladed rotors so that, in general, you will find them only on rotor systems of three blades or more. It is hazardous to generalize, but articulated systems tend to be mechanically complex and dynamically smooth, while offering moderate control authority. You have seen articulated rotors on the Hughes 500, and all Sikorsky aircraft.

Teetering rotors are all two-bladed. The blades are rigidly attached to the hub, but the hub itself is free to tilt and rock with respect to the rotor shaft. A teetering hinge permits seesaw motions in which one blade flaps up as the other flaps down. A rocking hinge allows feathering motions and there is no need for lead/lag hinges because those motions are absorbed through blade bending and gimbaling.

In general, teetering systems are mechanically simpler and dynamically less smooth than articulated ones, with somewhat less control authority and considerable control lag. Teetering rotors have been successfully employed on all commercial Bell helicopters.

Semi-rigid rotors come in several varieties and shades of meaning. In a semi-rigid rotor system the blades are rigidly connected to the rotor hub with flapping and lead/lag motions accomplished through the natural elasticity of the blades themselves. They are also called hingeless rotors, which is, in fact, an apt description.

TERMINOLOGY FOR THE FIXED-WINGER

Bell sometimes refers to its two-bladed teetering systems as being semi-rigid. That is technically correct because the individual blades do bend enough to perform some limited flapping and leading/lagging, although semi-rigidity is of minimal significance in teetering systems where the two blades are free to interact naturally.

Semi-rigid rotor systems of three or more blades are different. They are relative latecomers with excellent potential. These hingeless systems are less complex and invariably use fiberglass blades that are not life limited.

In general, semi-rigid rotor systems are simple and maintenance-free, exhibit a slightly harsher ride quality at maximum forward speeds and provide outstanding stability and control. You can see one on the BO-105.

Rigid rotors, truly rigid rotor systems, are still confined to research aircraft. One such aircraft is the Sikorsky ABC helicopter with counter-rotating rotors and blades stiff enough to walk on. Rigid rotors lend themselves to flight at higher speeds than normal helicopter rotor systems and so can be used on converter planes or rotorcraft with fixed wings to provide lift after takeoff.

Rigid rotors may have some commercial application but not in the foreseeable future.

A MATTER OF DEGREE

Despite those neat, theoretical categories, it is often difficult to precisely pigeon-hole a given rotor system. The Aérospatiale Gazelle employs a so-called semi-articulated rotor system in which flapping and lead/lag motions are shared between hinges and blade flexibility. The Aérospatiale AStar and Twin Dauphin rotor systems are called "articulated" although the fiberglass hinging used is quite similar in concept to hingeless, semi-rigid rotor types. The ultra-light HS-180 Hunter uses an unusual, articulated two-bladed rotor, and the possibilities go on and on.

Rotor system design is a rapidly changing art, although new types must still provide those three essential axes of motion: flapping, lead/lag and feathering. Articulated, teetering and semi-rigid types will continue to predominate, but hybrids will often provide unique and creative solutions to the same basic equation.

There are at least two other rotor components that may or may not be incorporated into any given design:

HELICOPTERS

Lead/lag dampers or *drag dampers* are often used on articulated rotor systems to limit lead/lag motions. These dampers are usually in the form of small hydraulic shock absorbers with some sort of oil reservoir. When present, you will find one damper for each blade, mounted at the blade root, or "cuff," as it is also called.

Droop stops are mechanical devices that prevent the blades of articulated rotors from drooping too close to the ground, or the helicopter structure itself, during start-up and shut-down. At normal rotor rpms, centrifugal weights remove the droop stops.

THE ROTOR CONTROLS

As mentioned, rotors are complex and in normal flight that whirling system is a dervish of bending, flapping, leading/lagging, rotating and tilting motions, largely controlled by two cockpit devices, the cyclic stick and the collective lever.

The *cyclic pitch control stick* can be thought of as simply the stick. It is used to pitch and roll the helicopter just like the stick in an airplane, although those attitude changes are accomplished in an entirely different fashion.

The cyclic stick is used to tilt the rotational plane of the rotor or "tip-path-plane," as it is called. When the cyclic is moved, the tip-path-plane tilts in the same direction, creating a horizontal thrust component that tilts the helicopter in pitch and/or roll. Thus they serve the purpose of the ailerons and elevator of an airplane. But they accomplish the same end in a much different and more complex way. The odd name "cyclic" refers to the method by which that rotor plane is tipped.

Consider a simple pitch-down maneuver, for example. When the cyclic is pushed forward, the rotor system is affected so that the pitch on each blade is sequentially increased as it passes the 180-degree position at the rear of the machine and decreased as it passes the zero-degree position ahead. This cyclical change in blade angle of attack causes an increase in lift at the rear and a decrease in lift forward. The result is that the entire rotor disc tips forward, eventually tipping the fuselage with it.

The cyclic pitch control stick induces cyclical changes in blade pitch that translate to those simple pitch and roll motions. You can call it the cyclic or the stick. It doesn't really matter.

The *collective pitch control lever* is the primary power control. It is normally located to the left of the pilot seat and hinged at the back.

TERMINOLOGY FOR THE FIXED-WINGER

The collective moves in a simple up-and-down motion to simultaneously and equally change the pitch on all rotor blades, thereby regulating rotor thrust or lift. The term collective refers to the collective, or mutual, effect of this control on each individual blade.

Piston-engine helicopters have a motorcycle-type twist throttle on the end of the collective lever which is used to adjust the manifold pressure and thus maintain a constant rotor rpm. Turbine-engine helicopters normally have no such engine control on the collective because turboshaft engines are internally governed at a constant speed. One result is that turbine-engine helicopters require far less power control and monitoring and, in that respect, are much easier to fly.

In summary, the cyclic changes blade pitch sequentially to provide pitch and roll control. The collective lever changes blade pitch simultaneously and equally to control the magnitude of the thrust delivered by the rotor system. In short, the collective controls lift.

THE MECHANICS OF CONTROL

Both accomplish those pitch changes through a device known as the "swash plate."

The *swash plate* assembly is located directly below the rotor. It is the device that transmits control inputs to the revolving rotor system. It consists of two discs through which the rotor mast passes. The lower disc is linked to the cyclic and collective controls and does not rotate. It is sometimes referred to as the "stationary star." The upper disc is linked to the rotor blade pitch horns and turns with the rotor. It is called the "rotating star." The two discs are separated by a bearing surface.

Torque to the rotor is transmitted by a shaft from the power train that runs through the swash plate assembly. Blade-pitch changes are transmitted through the swash plate itself.

Collective control inputs move the lower, stationary disc straight up or down, which forces the upper, rotating disc in the same direction to simultaneously and equally change the pitch of all blades for power changes.

Cyclic control inputs *tilt* the lower disc, which forces the upper disc to tilt in the same direction. When the swash plate is tilted, the pitch of each blade is changed sequentially to tilt the rotor tip-path-plane for pitch and/or roll control. The bearing minimizes friction between the upper disc (rotating star) and the lower disc (fixed star).

(In actual practice, cyclic control inputs are transmitted 90 degrees ahead of any given blade, to compensate for gyroscopic precession, but that is a simple design problem of minimal interest to the pilot.)

The *anti-torque rotor* is that little sideways propeller on the back. It performs two essential functions. It counteracts the enormous torque of the main rotor and provides normal directional control.

Think of the tail rotor as a constant-speed propeller whose pitch (and thereby thrust) is controlled by the rudder pedals. When tail-rotor thrust precisely equals main rotor torque, there is no yawing movement. When the rudder pedals are moved to reduce tail-rotor thrust, main rotor torque will rotate the fuselage. When tail-rotor thrust exceeds main-rotor torque, the fuselage will rotate in the opposite direction.

The Aérospatiale Gazelle and Dauphin models have a ducted-fan-type tail rotor, which is called a fenestron, after a French word for round window. The fenestron is an interesting innovation with some advantages, but in general it provides directional control just like a conventional tail rotor.

You should note that when the thrust, or lift, provided by the main rotor is reduced, the main-rotor torque is reduced and there is a reduced need for compensating side thrust from the tail rotor. Conversely, when the main rotor is called on to produce more lift, more thrust is necessary from the tail rotor. Thus, just when more engine power is needed for the main rotor, the tail rotor also must have more power. Therefore in certain conditions of maximum performance, particularly in a small helicopter, a turn can be made in one direction, but not the other because the tail rotor robs the main rotor of too much engine power. Also, in lifting off or hovering in certain wind conditions, the maximum thrust provided by the tail rotor is insufficient to counteract the main rotor thurst and an involuntary turn results.

That interaction becomes instinctive knowledge to a helicopter pilot after a time, but in the beginning of his training it makes learning interesting.

There are other terms associated with helicopter hardware and at least four bear mentioning:

Elastomeric bearings are not limited to helicopters, but they are frequently mentioned in helicopter sales literature.

Elastomers are substances composed of long, tangled chemical chains that harden to elastic, rubbery solids when heated. Elastomeric bearings are made by bonding elastomer material and metal in several sandwiched layers. These unusual bearings are comparatively inexpensive, reliable, maintenance-free, and corrosion resistant.

Stability augmentation systems (SAS) and *automatic flight-control systems* (AFCS) are integral parts of many new helicopters, but the terminology can be misleading. SAS is an electronic device that provides very-short-term rate damping. SAS has no exact counterpart in fixed-wing aircraft, but it more or less artificially does what dihedral and the fixed vertical and horizontal stabilizers do for an airplane aerodynamically. Since helicopters have little or no natural aerodynamic stability—depending on a number of design factors—SAS, as its name implies, augments the stability of the machine. AFCS is an electronic system that actually controls the helicopter, just as an autopilot flies an airplane.

The *transmission* transmits engine power to the main and tail rotors. It is a vital, dynamic component that requires periodic overhaul like a power-plant.

The *tail rotor gear box* or *90-degree gear box* is located at the tail rotor where it changes the plane of rotation of the tail rotor drive shaft 90 degrees to accommodate the side facing tail rotor.

New terminology appears every year as technology advances, although these basic items remain central to an understanding of helicopter hardware. Operational terminology is another whole subject.

THE HOVER

You already know that the collective control is used to increase or decrease the pitch of all blades equally and simultaneously as the primary means of lift control. Movements of it must be accompanied by changes in engine power as appropriate, either as the result of manual coordination (the usual case with piston power) or automatically (the usual case with turbine power).

The basic action required to lift off, then, is a gradual increase in collective pitch and power until thrust (lift) exceeds weight and the helicopter flies. At this point, several terms and definitions intrude:

271

Blade coning is the natural upward bending of the rotor blades into a conical pattern when they are producing lift. It is comparable to the upward bowing of a flexible wing in flight. Blade coning is limited by centrifugal force acting on the mass of the blades, augmented, usually, by weights at the blade tips. Obviously, the thin, limber blades of a helicopter are not by themselves capable of sustaining the weight of the machine when they are static. But in motion that centrifugal force adds tremendous rigidity to the blade assembly.

As rotor rpm decreases, the rigidity lessens and the coning increases. It can reach critical proportions if rotor rpm is allowed to drop substantially below the normal range. Therefore, the helicopter pilot flies with one eye on rotor rpm, just as a fixed-wing pilot flies with one eye on airspeed.

Ground effect in a helicopter is analogous to ground effect in a fixed-wing aircraft—but different. When a helicopter hovers low enough—within about one rotor diameter—the ground plane beneath it alters the vertical airflow component, which consequently affects the flow pattern through the rotor. In essence, the ground plane restricts downward flow through the rotor system and this—just as in a fixed-wing aircraft—enables the airfoil to produce more lift with less drag. Thus, the helicopter can always hover on less power in ground effect (IGE) than it can out of ground effect (OGE).

Ground effect is often compared to a bubble of high-pressure air. That's a useful analogy, but technically inaccurate. Nevertheless, when a helicopter is coaxed into forward flight out of an IGE hover, it often feels like it has fallen from a high-pressure bubble, especially at limiting weights, altitudes, or power.

Maximum weight or altitude for hover IGE is a prime measure of helicopter performance and those figures are invariably included in sales literature and flight manuals.

But note the IGE hover ceilings shown in flight manuals assume a flat, clean surface. One that is broken up by huge rocks, heavy vegetation or other irregularities will cause a decrease in the IGE hover performance.

Ground effect is inversely proportional to height above the surface, so it is very pronounced when a helicopter hovers scant inches off the ground and decreases with each succeeding foot of altitude. Flight-manual IGE hover charts usually stipulate the "skid height" to which they apply.

TERMINOLOGY FOR THE FIXED-WINGER

272

Hovering out of ground effect (OGE) is exactly what the name implies. It is hovering at any altitude above the point where the ground plane influences airflow through the rotor. In general, OGE hover is any altitude above about one rotor diameter. For any given weight, a helicopter will require significantly more power to hover OGE compared to IGE. Conversely, at maximum power and any given weight, the OGE hover ceiling will occur at some significantly lower density altitude than IGE hover.

Hover ceilings are limited primarily by the amount of power available to the main rotor. Add a bigger engine and you invariably increase hover ceilings. As a broad measure, OGE hover ceiling indicates the performance capability of a helicopter under the least favorable conditions, including no wind.

Translational lift will add to the efficiency of the rotor system and reduce the power required as the helicopter begins to move forward, creating a horizontal airflow. In other words, a rotor system produces more lift for a given power setting and blade pitch angle in forward flight, or in a breeze, because it has a greater inflow of air per unit of time.

The effect is so pronounced a helicopter pilot knows instinctively that he will need to reduce collective gradually as speed increases up to some moderate level. From that point he will need to add more collective and power to develop the thrust vector necessary for high speed. Maximum power requirements always occur in the hover and at high speed.

One other obvious effect of translational lift is the onset of noticeable buffeting as the flow state of the rotor changes from hover to forward flight. When the prevailing wind velocity is just right, translational buffet can be constant during the hover. In general terms, translational lift occurs at about 15 to 20 knots' airspeed, causing some degree of buffeting as the helicopter is accelerated from or slowed to a zero-airspeed hover.

In summary, OGE hover in a no-wind condition will require the most power. Any amount of wind will proportionately increase rotor efficiency due to translational life and reduce power requirements. Also, any reduction in hover height above the ground below about one rotor diameter will do the same. You should also know that the effects of translational lift and ground effect are independent although complementary. Hover OGE in a stiff breeze can require only a modest percentage of the power used in IGE hover in still air.

HELICOPTERS

DISSYMMETRY OF LIFT

In a hover, particularly in calm air, the rotor enjoys a steady state in which airflow over the blade is a simple function of the rpm of the rotor. Once the helicopter moves forward into that air, through translational lift, the rotor system dynamics become signficantly more complex.

Advancing blades, retreating blades is an important concept in understanding the dynamics of forward flight in a rotor-wing aircraft. During each complete revolution, a given blade will spend half of its journey advancing in the same direction as the helicopter and half retreating in the opposite direction. If the rotor turns counterclockwise, as most do, the blades will be advancing on the right side and retreating on the left. This advancing/retreating blade fact-of-life creates a classic helicopter idiosyncrasy called dissymmetry of lift.

The circle described by the top of the blades is called the *rotor disc*. In a no-wind hover, each rotor blade produces a fixed and constant lift as it rotates around the rotor disc so that lift distribution is balanced.

As the rotor system begins to move forward into the air, however, the relative wind over any given blade becomes a combination of rotational speed and the forward movement of the helicopter. At the three o'clock position, for instance, the speed of the advancing blade is the sum of its rotational speed and the airspeed of the helicopter. At the nine o'clock position, the same blade will be retreating so that its speed will then be the *difference* between rotational speed and airspeed.

To illustrate further, the typical rotor-tip speed of a small helicopter at zero airspeed might be 350 knots. At 100 knots airspeed, the relative wind over the advancing blade tip will be 450 knots. But the relative speed of the retreating tip will drop to 250 knots.

As airspeed increases, the advancing blade experiences progressively more airspeed and the retreating blade progressively less, assuming the rpm is constant. If no compensation were made, the advancing half of the rotor disc would produce significantly more lift with airspeed increases than the retreating half and the helicopter would eventually roll out of control at some critical airspeed.

Some very early experimental helicopters did just that.

There is much disagreement on this point, but you can believe that the dissymmetry problem is resolved in two ways.

TERMINOLOGY FOR THE FIXED-WINGER

First, the cyclic control is simply displaced to the right against the rolling tendency. This tilts the swash plate in such a way that the feather—in effect the angle of attack—of the advancing blade is *decreased* while the feather of the retreating blade is *increased*. This, obviously, equalizes the lift generated by the faster-moving advancing blade relative to the slower-moving retreating blade.

Test pilots tell us this is, in fact, how dissymmetry is overcome. However, another factor enters in and it is the usual textbook explanation of why the helicopter doesn't roll over.

BLADE FLAPPING

Flapping is the vertical motion of the blade about the flapping hinge. As the blade rotates in forward flight, through the advancing half of its cycle, the increased airspeed results in extra lift, which causes the blade to flap up. That up-flap effectively decreases the angle of attack of the blade by altering the direction of the relative wind. As the blade turns the corner to become a retreating blade, the airflow over it is reduced so that it produces less lift and consequently flaps down. Down-flapping increases the effective angle of attack of the blade.

This automatic sequence of up-flap on the advancing side and down-flap on the retreating side tends to equalize lift across the rotor disc. This flapping action is designed into the rotor by virtue of flapping or teetering hinges or, in the case of hingeless systems, by the natural elasticity of the blades.

RETREATING-BLADE STALL

At some maximum airspeed, however, the helicopter will begin to encounter another problem. In your airplane, the minimum speed is limited by wing stall. In a helicopter, maximum airspeed is limited by blade stall.

As the helicopter flies faster, the effective angle of attack of the retreating blade will increase more and more to balance the lift on the advancing blade. At normal cruising speed, the advancing blade produces lift with small angles of attack at relatively high airspeed, while the retreating blade operates at very high angles of attack to produce an equal amount of lift at a reduced airspeed.

At some limiting point, the angle of attack on the retreating blade will exceed the stalling angle of attack for the airfoil section at that airspeed and it will shed lift. The onset of the stall will be at the tip and will progress inward as the airspeed increases.

At high indicated airspeeds, blade-tip stall will be felt as an obvious heavy vibration in the controls and airframe as each tip sequentially stalls on the retreating side. This buffeting can be eliminated by simply reducing collective pitch, which reduces all blade pitch angles and relieves the stall.

Note that the tip stall may be aggravated by a cyclic input to the right although the consequent roll to the right may somewhat relieve it.

In an extreme condition, the loss of lift from blade stall at the nine o'clock position will result in large down-flap displacement at six o'clock, causing the fuselage to pitch up. Really severe retreating blade stall could include strong left-rolling forces, but there is no reason ever to encounter that degree of blade stall at all. Retreating blade stall provides a natural upper limit to the speed of the helicopter, which is clearly signalled by gradually increasing vibration.

One note on blade-tip stall and top speed. Helicopter manufacturers always list some cruising airspeed that is quite naturally limited by the onset of blade stall. Since blade stall is aggravated by gross weight, blade condition, G loading, rpm, high-density altitudes and turbulent air, cruising speed claims are often optimistic expressions based on ideal test conditions. The maximum comfortable speed is often as much as ten percent less than the listed top speed, especially under normal conditions of blade erosion and rotor mistrack.

AUTOROTATIONS

Autorotation is simply a power-off descent. As the name implies, the rotor turns itself during this maneuver, or more accurately, is driven by the aerodynamic affects of the relative wind. In powered flight, airflow is downward through the rotor. In autorotation the airflow is reversed and flows upward, actually driving the rotor. Demonstration of adequate autorotative capability is required of all helicopters by FAA certification standards.

Autorotation is normally associated with engine failure in single-engine helicopters and that, in fact, is the most common reason for performing the maneuver—simulated engine failures for training purposes included, of course.

When the engine stops supplying power to the main rotor, the pilot immediately reduces collective pitch to the minimum and begins a reduction to best glide speed, usually about half of cruise speed.

In a stabilized autorotation, the rotor disc can be broadly separated into three regions. The inner 25 percent of the rotor diameter will be stalled, creating a small amount of drag that tends to slow rotor rpm. The next 50 percent, referred to as the driving region, produces the rotational forces that keep the rotor turning. The outer 25 percent, or "driven region," produces lift, which slows the rate of descent.

The autorotation maneuver is a classic exercise in energy management. As the helicopter approaches the ground at a normal autorotation rate of descent of 1,200 to 1,800 fpm, the pilot flares steeply, a maneuver that breaks the rate of descent while increasing rotor rpm. Then, using the considerable inertia of those rotating airfoils, he cushions the landing by increasing collective. Rotor rpm rapidly decays during the final touchdown, but there is more than enough to control the helicopter to a safe and gentle landing.

Airplane pilots often find their first ride through an autorotation to be somewhat alarming because of the steep descent and rapid flare maneuver at the bottom. But the fact is that autorotations are not difficult. A competent pilot can autorotate with precision and finesse to any designated area within gliding distance.

POWER SETTLING

Power settling is a helicopter phenomenon somewhat similar to flying behind the power curve in a jet-powered airplane. During a descent at high blade-pitch angles and high power settings, but at a very low forward speed, the helicopter can descend into its own downwash, which will effectively stall the inboard section of the rotor. Adding power will serve only to aggravate the situation and increase the descent rate because the rotor is largely recirculating disturbed air.

Power settling is avoided by maintaining some minimum forward speed and limiting the power and/or rate of descent.

Power settling can result from an attempted long vertical descent, which is one reason helicopter pilots do not eagerly respond

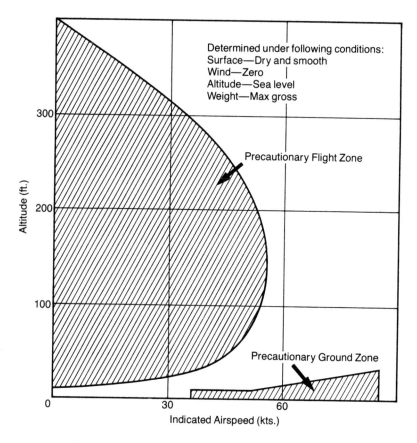

Determined under following conditions:
Surface—Dry and smooth
Wind—Zero
Altitude—Sea level
Weight—Max gross

to requests to land in confined areas surrounded by steep obstacles.

THE HEIGHT/VELOCITY CURVE

Helicopters are VTOL (Vertical Takeoff and Landing) aircraft, but the "V" is often compromised by operational considerations beyond even power settling. Some of those considerations are combined into a single flight-manual chart.

The Height/Velocity diagram above, or H/V for short, identifies those portions of the flight envelope from which a safe landing cannot be made after a sudden power failure. The H/V diagram is also known as the Altitude Versus Airspeed Chart or the Dead Man's Curve. The chart above is an example.

The H/V diagram takes into account the airspeed and/or altitude necessary to transition from normal flight to an autorotation.

278

Operation within the shaded area does not provide enough air-speed or altitude for a safe transition and autorotation following total engine failure.

The safety-conscious pilot will plan takeoffs and landings so as to avoid either shaded area. When operational imperatives require him to fly within the shaded areas, he knows that a safe auto-rotation will not be possible. It's a little like flying below V_{MC} in a multi-engine airplane. It can, and sometimes must, be done, but there is no margin for safety.

HELICOPTER BALANCED-FIELD LENGTH?

Because of the H/V ratio requirements, helicopters have a safe-field-length requirement not unlike the balanced-field-length re-quirement for Part 25 fixed-wing aircraft.

We routinely think of helicopters as vehicles settling down onto tennis courts in the midst of skyscrapers. They do have the capa-bility of doing that, but the prudent helicopter pilot avoids doing it often. He doesn't like to fly into such a landing area because of the potential for a power-settling incident as explained above. But there's also potential for an accident in the departure, and it springs from the H/V chart.

If the pilot is going to avoid those dreaded shaded areas, he needs some distance in which to lift off and accelerate through that window between the precautionary flight zone and the pre-cautionary ground zone. For the corporate helicopter pilot, that's a crucial distance. If an accident should occur because the appro-priate distance wasn't available, the liability risk would be great. Yet few, very few, helicopter manufacturers include take-off dis-tance data in their operations manuals. This is a serious failing because that distance can be considerable.

The example, notice, shows that 550 horizontal feet of unob-structed field length are needed for a safe departure over a 50-foot obstacle on a standard day at 1500 feet msl. That doesn't have to be level, flat ground—when departing from a rooftop, for example, the precautionary ground zone may drop out of the H/V chart in three or four feet of horizontal distance—but the helicopter still requires some minimum distance in which to accelerate to a safe autorotational speed before climbing steeply away.

In the case of a departure from a rooftop or platform, some of that distance can be traded off for the altitude advantage. Again, manufacturers do not give a chart for computing that safety tradeoff, but the H/V curve for a particular helicopter will give the pilot a clue. Using the H/V illustration shown here, for example, if the precipice is 400 feet high or more, no takeoff distance is needed because the liftoff is above all shaded areas. On the other hand, if the altitude available for transitioning into autorotation after leaving the roof is only 200 feet, a safe takeoff demands enough distance to accelerate to about 55 knots before flying over the edge.

Using the H/V chart in this way doesn't guarantee safety, but it does give the pilot some idea of the risk involved in departures from raised areas.

VIBRATION TERMINOLOGY

Helicopter pilots develop a sensitivity to vibration that is akin to a musician's ear. Vibrations are often the first indication of trouble. Basically there are three types:

Low-frequency vibrations are always associated with the main rotor and are usually identified as one-per-rev, two-per-rev, three-per-rev and so on based on main rotor rpms. Depending on frequency, intensity and orientation, the pilot may be able to diagnose the problem with great accuracy. Helicopter mechanics are accustomed to logbook comments that reference such vibrations.

Medium-frequency vibrations are usually associated with the tail rotor and can often be felt through the rudder pedals. Tail rotors are very delicately balanced components and failure is a serious emergency. A medium-frequency vibration may telegraph the problem before any failure occurs, so when one is felt, a precautionary landing is in order.

High-frequency vibrations are normally associated with the power plant. The effectiveness of high-frequency vibrations as malfunction indicators has greatly diminished with the advent of turbine power plants. Still, high-frequency vibrations, when present, are an indication of engine abnormality.

Ground resonance is a self-exciting airframe vibration produced by rough ground contact which, in its worst state, can demolish a helicopter in seconds.

Ground resonance is generally limited to older design, fully articulated rotor systems. It begins when a shock to the landing gear is transmitted through the fuselage to the rotor head. If that shock is great enough, it will displace one or more of the blades, momentarily unbalancing the head. That momentary imbalance now pulses down through the fuselage and is reflected back when the landing gear makes ground contact again.

Resonance will be eliminated by lifting the helicopter into a hover and breaking all ground contact. Today, ground resonance is more of a cruel memory than an actual threat because newer landing gear and rotor designs have essentially eliminated the risk.

Helicopters are as different from airplanes as boats are from cars. The operation, price, design and terminology for rotorcraft are each unique.

You may never have the opportunity or even the inclination to fly a helicopter, but you cannot avoid them altogether. We all share the same airspace and serve a single, basic air-transportation need. Helicopters are an ever-larger part of that system and a little understanding is necessary for friendly, thoughtful coexistence.

END OF THE YEAR, END OF A BOOK

VII

Christmas is a special time in most American homes. It marks the birth of Jesus and the end of the year. It is a time to remember, reassess and reflect. It is a particularly meaningful time for families.

We had a Christmas celebration each year on the "Crossfeed" page for our family of readers. There were memories, laughs, and lots of presents. And just as these pages closed the door on individual years, they also seem appropriate to close the curtain on this book.

Christmas Thoughts

44

I wanted to buy you all a Christmas present, but I ran into difficulty in locating most of the items. However, considering it's the thought that counts rather than the actual gift, the following list will show what I had in mind.

First, I wanted to send along some realistic salaries and working conditions to that majority of business aviation people who labor for inappropriate or inadequate wages. I mean those construction company pilots earning less than unskilled laborers, and mature executive pilots whose compensation barely exceeds that of a management trainee. I mean also the mechanics who carry total maintenance responsibility for million-dollar planes on janitor's wages. In fact, this package was to be wrapped in festive fringe benefits and tied with strong job security. The tags are all made out, so as soon as I can locate those items they will be sent right along.

Next on my list was a large box of nothing. Nothing, that is, in the way of governmental obfuscation, strangulation and pointless administration. I had hoped that this package would be as big as the new year—and as empty as a bureaucratic memo. Unfortunately, this particular item has been back-ordered since the New Deal, and very few people can even remember what it looks like. I'll keep trying, but quite frankly there is little hope of delivery in this century.

One little trinket I had hoped to put in your stocking is a new type of sunshield. I'm not really sure what it looks like, but it will be light, flexible, and at least as adaptable as those Jepp charts we're all using now. It will be transparent, too, so we can see each other in severe VFR conditions. Meanwhile, pick up another roll of tape to stick those Jepps in all the sunny places.

END OF THE YEAR, END OF A BOOK

I had planned a unique gift for those poor souls who earn their pay in the low-altitude sector. They were supposed to get an ombudsman in each Air Route Traffic Control Center, a guy who would make sure that their requirements were taken seriously. That ombudsman would just stand around until some poor soul in a Cherokee was issued a 9,000-foot clearance for a 30-mile leg. Right then, he'd scream and yell and carry on until some more realistic altitude was issued. Or, when the radar controller would say that you can't get there from here because of all the airline departures, the ombudsman would leap four feet off the floor and threaten to flatten every tire in the parking lot. The prototype looked like John Wayne, but it never went into production and the entire project has been shelved.

For high-altitude pilots, I wanted to arrange a new international treaty. That agreement between Canada and the United States would have settled the nomenclature for high-altitude airways. From Michigan to New York, it would no longer be necessary to file J547-HL547-J547. Just one entry would suffice, and you could donate the money saved on pencil lead to your favorite charity. I called the Secretary of State several times, but he was always tied up with some Middle East crisis. Maybe we need an ombudsman at the State Department too.

One thing I had in mind for everybody was a whole year free of revisions. Nothing. No new charts, maps, plates or avigation pages. In fact, this one was a sure bet in light of the paper shortage. But I ran into an unexpected snag: my kids needed the old maps for schoolbook covers.

I looked and looked for a bleep expunger. That device would eliminate the 150-dB BLEEP following every ATIS transmission, making raw meat out of my eardrum. The advanced model would have a second mode to erase the Morse Code (CW) identifier in those cases where it is broadcast simultaneously with the verbal ATIS information, since you really can't hear either one satisfactorily. I was able to find whoopers and squawkers, but there just aren't any certified bleep expungers.

One grandiose plan involved the recruitment of an army of craftsmen to service your airframes, engines and avionics. These impeccable workmen would actually have charged rates somewhat above the norm, but their work would never, never include the shoddy treatment we are gradually coming to take for granted: greasy handprints all over the airplane, scratched and dented engine cowls and wrinkles on the stabilizer skin from engine oil being thrown on it to lift the nosewheel off the ground. But most of all, my army would have featured precision, professional results with courteous treatment for the customer with a grievance. Even the uniform looked great. It had a little emblem that said "Pride. Quality. Value."

Unfortunately, I didn't enlist a single man. Everyone I tried was too busy to answer the phone.

Helicopter operators were supposed to get an IFR package at a price not exceeding that of gold bullion on an equal-weight basis. This article would even have fit into existing machines on a strap-on basis and satisfied the *real* needs of these unique aircraft. The selection was extremely limited, but it's just possible that things will be somewhat better next year.

I had one other present in mind for my rotary friends. That was a sound-proofing package that reduced interior noise to the level of a Sherman tank. I tried to contact the outfit in Hollywood with those whisper-quiet helicopters I see in movies, but no one knew who they were.

They say that books always make a nice gift, so I wanted to wrap up some new flight manuals. These volumes were to have a standard format, good indexing and realistic performance tables in a new, small size. The General Aviation Manufacturers Association has started a project to create manuals like the ones I had in mind, but it still has a long way to go before the manuals it produces can be considered gifts.

Finally, I let out a contract to destroy that enormous underground pipeline that distributes war-surplus coffee to all the aviation caterers. Unfortunately, even though that conspiracy is well-known, the pipeline could not be found. As a substitute we have formulated this page from pure antacid so that it may be consumed after your next gut-wrenching cup.

Well, that's about it. At least my heart is in the right place.

More Christmas Thoughts

45

Some of you may remember my Christmas shopping list of last year. Included in that list were several items I had wanted to send along to each of you. Unfortunately, I was frustrated at every turn and as a result none of those gifts were ever actually sent.

This year I resolved that things would be different. My shopping spree begin in early spring and has included hundreds of ideas, ten months of searching and 300,000 miles of travel. Regrettably, I was not very successful in locating the items on my final list, so don't bother to watch for the mailman, expecting to receive my package. Sorry, but just so you know where my heart is, here's what I had in mind.

The first item was to have been a gift certificate for one thousand. Not limited to any one specific thing, mind you, but one thousand of whatever you need most. Here's how it worked:

You get a beautiful, inscribed certificate that fits right into your Jepp binder. On the face it says, "Redeemable on demand for 1,000." Now, anytime during the year you need a thou, you just surrender the coupon. Like if you arrive back at home base from a tough trip and the RVR is below minimums. You just inform Approach Control that you are exercising your coupon and the RVR goes up 1,000 feet. Or maybe the weatherman blows the en route wind forecast again, and the 80 knots of tailwind turn into 30 knots of headwind. You just wave that coupon at the fuel gauge and it promptly increases by 1,000 pounds.

Actually, the possible uses would have been almost endless. An extra 1,000 feet of altitude to top the weather. An extra 1,000 feet of ceiling to go VFR. Even an extra 1,000 feet of runway if that was your choice. Then, if your planning were so superb the coupon never got cashed in, you could trade it in for a Lucky-Thousand Club lapel pin ($2.95 to cover air freight).

END OF THE YEAR, END OF A BOOK

Actually, I had the certificates all printed and addressed but the FAA, FCC, NTSB, DOT, FEA, CAB and all those government bureaucrats were concerned about regulations and violations, and somebody stonewalled the whole thing. As it turns out, I'm using the certificates to train my puppy.

Next on my list was a whole bunch of wind socks. Many of you will not remember those big cloth wind indicators because someone apparently figured a long time ago that wind socks were not compatible with jet sophistication. You hardly ever see them anymore.

In reality they were very nice to have because right there at the end of the runway you could actually see what the wind was doing without relying on some flaky electronic gizmo in the tower three miles away. I had planned for volunteers to knit them in nice argyle patterns, but I hit a dead end. No one seems to know how to knit argyle wind socks anymore.

I tried to establish a tax-exempt research foundation to examine the question of knobs. You know what I mean: radio knobs, biffy door handles, suitcase grips, that sort of thing. I even found a guy with several advanced degrees in knob knowledge to head up the program. His first project was to investigate why the most critical knob always fails first.

As it turns out, the Memorial Institute for Knob Research was sabotaged by the international cartel that produces all those slipping, binding, failing models, which are standard in aviation. Too bad, but you can't fight City Hall.

One little trinket I had hoped to put in your stocking was a missed approach. Not the actual event, you understand, but a new procedure that would standardize that semi-emergency in some reasonable, orderly fashion. Stuff like "Climb straight ahead to 600 feet, then right climbing turn to 025 degrees, continue climb to 3400 feet via outbound XYZ VOR 094 R to intercept ABC VOR 351 degrees R to PANIC intersection to hold northeast with left turns" would be changed to "climb straight ahead to 3,000 feet and remain on tower frequency for radar vectors."

Everybody liked the concept except the President's chief economic adviser. He said that the idea would increase unemployment by at least one percent from the layoffs of flight-procedures experts and chartmakers. Since this was an election year, you can guess the outcome.

Jet pilots were supposed to get a special goody. It would have been a little black box designed to jam the ATC mental telepathy that enables the controller to detect your throttle position. That mental Throttle Position Indicator (TPI) is what tells the controller when you have leveled off and adjusted the throttles; it's his signal that he should immediately issue another climb or descent clearance. With the controllers' TPI circuit jammed with the black box I had planned for you, passengers could enjoy the peace and security of one steady thrust setting all the way up and down.

The TPI jammer was almost into production when I discovered that the prototype opened electric garage doors in a 200-mile radius.

Books are always appropriate so I looked for a nice edition of FAR 91. My copy would have been organized in an unusual fashion, with similar subjects grouped together. As it turns out, FAR 91 is supposed to be confusing, and in fact has been specifically designed as the only perfectly random assembly of unassociated (and irrelevant) ideas. When you think about it, how else could you group emergency exits, truth in leasing, and sonic booms on the same page?

Actually, I had a bunch of other stuff in mind as well. Instrument pilots were supposed to get a new set of en route charts with all critical frequencies moved away from the folds, where they are so easily obliterated. And everybody was due for a few incidentals: circuit breakers that don't break your fingernails; one free visit to a chiropractor to correct back problems from crummy cockpit seats; and a fuel drain that doesn't empty into your shirtsleeve. I'll try again for Christmas next year, but frankly I'm getting discouraged.

Anyway, have a real nice Christmas. And while you're at it, don't forget whose birthday it is.

Merry Christmas, 46
Happy New Year

It's hard to believe that Christmas is here again already and as usual I had a hard time with my shopping. I know that you can appreciate how difficult it is to find presents for my friends in the aviation community. I mean, everyone knows that people who work with airplanes make more money than they can possibly spend, so I have tried to find little things that will brighten your days in the coming year.

One of the more unusual items I came across was the Electric Grandmother. This amazing bionic contraption was designed to replace all those food machines that now dispense cold soup, warm milk and stale candy bars. When you drop your money into the E.G. (Mark I), a real, simulated grandmother gets out of her rocker and whips up fresh-brewed coffee, home-made fried chicken and deep-dish apple pie. While you eat, she nods understandingly as you tell her about the lousy weather, lousy airplane and lousy captain. When you're done, she cleans up the mess and automatically returns to a standby mode for the next customer. Unfortunately the airlines got there first and bought up the first five years' production for use as flight attendants. Some of them are in operation already and the only difference you can see is the improved service.

Along the same lines, I had hoped to develop an improved coffee. This unique product would have been made from an ancient recipe that calls for pure water, fresh coffee grounds and a clean brewing device. The end product would have been clear and fresh with a deep, rich smell and taste. I discovered at the last minute that an obscure section of the FARs requires all aviation coffee to meet the Mil Specs for hydraulic fluid and/or paint stripper. Naturally, I couldn't tamper with the law.

293

Helicopter and light plane pilots were supposed to get a turbulence control switch, labeled, "Seat Belt Sign—On or Off." The installation was to have been really easy because the switch wouldn't be wired to anything and could be placed anywhere in the cockpit. When you encountered rough air, you would just move the switch to "Seat Belt Sign—On" and the choppiness would stop immediately. The pilots of large aircraft have been doing it for years and they know how well it works.

One little goody I had lined up for all the cockpit types was an entirely new ATIS format broadcast in English. The idea was to provide useful information like departure and arrival runways, braking action and the weather trend in simple, slow, Midwestern English.

At the last minute the State Department intervened to preserve those traditional broadcasts being done under the cultural exchange program by Rumanian auctioneers.

I looked all over for a special type of oxygen mask. This soft, lightweight device would gently slide over the head without knocking your glasses into the rudder pedal well. Right inside it would be a microphone capable of transmitting the human voice with no more than 50 percent distortion. Finally, the oxygen selector would have been a three-position switch for apple-pie, prime rib or mint flavor. I'm still looking, but in the meantime, don't hold your breath.

I watched the TV news the other night and saw one of those Ted Baxter-types using a cordless lapel microphone to broadcast from the production floor of a boiler factory and that gave me another idea. Why not certify that kind of microphone for cockpit use and get rid of all the wires and brackets and boom mikes and hangers? As it turns out, the Federal Office of Aviation Image insists on real microphones because of their strong visual appeal to the general public. People just don't believe that pilots can talk without a hand mike. If you wish, you can lodge a complaint with Commissioner S. Canyon in Hollywood.

Actually I spent a great deal of time designing a new weather prognosis terminology. This highly accurate wording would have been used to replace such routine surface prognosis language as: "cut-off low mvs into srn Plto stag PVA assocd with trof generates upwd VV and high RH. Batrop gvs suppt to this fast." My version would have said something like "Rain in Los Angeles, fog in San Francisco, and clear in Denver."

MERRY CHRISTMAS, HAPPY NEW YEAR

I showed it to several meteorologists and none of them could understand it, so I scrapped the program.

Last year we developed the One-Thousand Coupon, redeemable, when needed, for 1,000 feet of altitude, RVR, runway, or pounds of fuel. This year I hope to be able to offer five Truth Coupons which can be used at your discretion in any aviation-related incident.

Suppose, for example, that Flight Service lists the destination weather as 4,000 broken and ten miles visibility with smooth air en route. When the trip is unusually important, you whip out a Truth Coupon and redeem it for the true weather situation. Immediately, the briefer will be compelled to tell you that your destination is covered with 32 inches of snow and that the only smooth air en route is the eye of hurricane Samantha. The coupons can be used for EACs, job interviews, aircraft purchases or during happy hour.

The post office said they would be delivered within three weeks. Maybe I should have used a coupon on *them.*

All year long I've been working on a new suitcase design to hold all the other presents. This neat little case will have steel hinges certified for 1,000 openings. Instead of those flashy little latches, it will have a piece of rope to tie it shut so you won't dump all your clean clothes into the puddle of avgas on the ramp. It will be small enough to fit in your baggage compartment, large enough to carry all your stuff and light as a feather. I'm still working.

Actually, I'm not too disappointed that my gifts didn't work out. They could never have competed with the real gift of Christmas anyway.

Quiz   47

Before *Crossfeed* ends, this is a good time for a quiz. First, the ground rules:

- Flight examiners, FAA inspectors and PPEs are forbidden to use these questions except among themselves.
- Chief pilots must post their own grades promptly and honestly.

THE TEST

1. When the Pitot tube freezes over in a climb, airspeed indications will:
   (a) Increase
   (b) Decrease
   (c) Remain constant

2. The lowest altitude at which a pilot may experience hypoxia is:
   (a) Sea level
   (b) 5000 feet
   (c) 10,000 feet

3. The lowest speed at which an airplane may hydroplane is primarily a function of:
   (a) Aircraft weight
   (b) Tire size
   (c) Tire pressure

4. On a nonprecision approach (NDB, VOR, back-course localizer) in IMC and before reaching MDA you are instructed to go around. You should begin the published missed-approach procedure:

(a) Immediately
(b) Upon reaching the minimum descent altitude
(c) At the MAP

5. You touch down at 100 knots. Some specific amount of total braking energy will be required to reduce speed from 100 knots to zero. What percentage of that total braking will be required to decelerate from 50 knots to zero?
   (a) 25
   (b) 50
   (c) 75

6. FAR 91.70 restricts airspeed to:
   (a) 250 KIAS, at or below 10,000 feet
   (b) 750 KIAS below 10,000 feet

7. For a given aircraft, added weight will affect the descent profile as follows:
   (a) No difference
   (b) Added weight will steepen the descent angle
   (c) Added weight will reduce the descent angle

8. Hail may be encountered:
   (a) In a thunderstorm
   (b) Below a thunderstorm
   (c) Within five miles of a thunderstorm
   (d) All of the above

9. A strong crosswind during landing will:
   (a) Decrease stopping distance
   (b) Increase stopping distance
   (c) Neither of the above

10. The notation "VIRGA" after a weather sequence report indicates:
    (a) A type of rain shower
    (b) Women pilots operating in the area
    (c) A type of cloud formation

11. You are making an ILS (back course) Runway 22 approach to Minneapolis-St. Paul International Airport. The Final Approach Fix is the SNELL BCM marker. When you cross SNELL, which marker beacon light will illuminate?
    (a) White
    (b) Yellow
    (c) Blue

12. At higher altitudes your altimeter must be set to 29.92. In the United States what is the lowest usable flight level with that setting?
    (a) FL 180
    (b) Above FL 180
    (c) Depends on surface pressure.

13. You are approaching Miami on a clear day and the DME takes an excessive amount of time to lock in. You should:
    (a) Be patient
    (b) Take it in for service
    (c) Tell ATC

14. One of your passengers appears ill and is observed to be breathing rapidly. She complains of light-headedness and a tingling sensation in the fingers, and you can see that her hands are cramped into a claw position. Her pulse is much faster than normal. You should:
    (a) Attempt artificial respiration
    (b) Administer oxygen
    (c) Apply an oxygen mask without administering oxygen

15. Mountain wave is the most intense and destructive form of nonconvective turbulence. The prime active areas in the United States are:
    (a) West of Denver, Fort Collins and Crazy Woman
    (b) West of Mono Lake, Cimarron and Las Vegas
    (c) East of Helena, Montana, and Alamosa, Colorado
    (d) All of the above

16. As you climb your jet to cruising altitude, you know that if one pilot leaves his station while the aircraft is above a specified flight level the remaining pilot will have to wear his oxygen mask until the other returns. That flight level is:
    (a) FL 410
    (b) FL 350
    (c) FL 250

17. Tower delays your departure behind a "heavy jet." You know by that designation that the preceding airplane:
    (a) Weighs over 300,000 pounds
    (b) Is certificated for an allowable takeoff weight over 300,000
    (c) Is loaded to max weight for the prevailing conditions

18. You are cleared for a visual approach to a runway served by an ILS and a VASI. Your glidepath during the approach:

(a) Is entirely up to your own discretion

(b) Must be at or above the ILS or VASI glideslope if those facilities are operating

19. You have been cleared to a fix with several crossing restrictions. Your clearance is amended to a lower altitude with no mention of the previous restrictions. Therefore:

(a) The previous crossing restrictions no longer apply

(b) You must conform to all previously issued restrictions

(c) You must clarify this point with ATC

20. When operating behind a large transport, which wind condition would result in the most persistent runway turbulence?

(a) Calm wind

(b) A direct headwind

(c) A 5-knot crosswind component

THE ANSWERS

1. (a) *Airspeed indications will increase.* If the Pitot tube becomes completely blocked with ice, trapped air in the system will expand with altitude, causing the indicated airspeed to increase. It's possible for a Pitot to freeze with the drain hole still open, in which case IAS would probably bleed off to zero. Pitot ice can actually lead to a number of different problems, so keep the heat on.

2. (a) *Sea level.* Medically speaking, hypoxia is insufficient oxygen in the blood. In aviation we tend to associate hypoxia with altitude, but it's entirely possible, in an extreme case, to be affected at sea level.

3. (c) *Tire pressure.* Actually, there are three different kinds of hydroplaning, but the most common one is dynamic hydroplaning, in which a wedge of water actually lifts the tire out of contact with the pavement. Dynamic hydroplaning may occur at any speed greater than nine times the square root of the tire pressure.

4. (c) *At the MAP.* Obstacle clearance and airspace allocation is predicated on missed approach maneuvers from the MAP. Earlier initiation of the missed approach compromises those clearances, although it is perfectly legal to begin climb prior to reaching the MAP.

5. (a) *25 percent.* Kinetic energy is a function of the square of the velocity. Thus, when you halve the speed, you reduce kinetic energy by four times. Early, judicious use of speed brakes, reverse thrust, and lift dump will disproportionately reduce the required braking.

6. (b) *250 KIAS below 10,000 feet.* The speed limit applies to *indicated* airspeed *below* 10,000 feet. Be aware that jet traffic often operates at up to 350 KIAS (about 400 KTAS) *at* 10,000 feet.

7. (c) *Added weight will reduce the descent angle.* It doesn't seem right, but your airplane will require more distance to descend when it is heavier, so plan those profile descents accordingly.

8. (d) *All of the above.* Hail is often thrown out of a thunderstorm by extreme updrafts and it can fall as much as five miles away.

9. (b) *Increase stopping distance.* Race-car drivers know that cornering forces on a tire reduces maximum available braking and during a crosswind roll-out, aircraft tires also sustain enough cornering forces to detract from maximum braking.

10. (a) *A type of rain shower.* VIRGA is rain falling from a cloud and evaporating before it reaches the ground. It is indicative of strong turbulence at lower levels.

11. (a) *White.* Back Course Markers are a rather new phenomenon. So far there are only a handful in the United States.

12. (c) *It depends on surface pressure.* A hair-splitting question, but one that emphasizes the safety implications of that altitude. The lowest usable flight level is never below FL 180, but it will be higher if the altimeter setting in the area is below 29.92.

13. (a) *Be patient.* The problem is probably saturation.

14. (c) *Apply an oxygen mask without actually administering oxygen.* Your passenger is likely suffering from hyperventilation and needs to restore the normal carbon-dioxide level in her lungs. An unconnected oxygen mask, or even a paper bag, placed over her mouth and nose will force her to inhale expired CO_2 and restore normal levels.

15. (d) *All of the above.* These eight locations are the prime areas for mountain-wave turbulence in the United States.

16. *(b) or (c), depending on your type of operation.* If you operate under FAR Part 91, the correct answer is (b), FL 350. If you operate under Parts 121 or 135 the correct answer is (c), FL 250.

17. *(b) Is certificated for an allowable takeoff weight over 300,000 pounds.* The "heavy jet" designation does not apply to weight at the time of the encounter. You could be held behind a DC-8 departing at 160,000 pounds or a 747 at 800,000 pounds.

18. *(b) Must be at or above the ILS or VASI glideslope if these facilities are operating.* See Part 91.87(d).

19. *(a) The previous crossing restrictions no longer apply.* If an amended clearance is issued and parts of a previous clearance are not reissued, the omitted portions no longer apply. This is true of routings and altitudes only. Airspeed restrictions need not be reissued. If in doubt, ask.

20. *(c) A five-knot crosswind component.* Wake vortices move laterally across the surface at about five knots. Therefore a five-knot crosswind component will tend to hold one vortex fixed over the runway.

The Christmas season ends with a look to the future, a new year. *Crossfeed* ends with best wishes for clear sky and tail winds on every leg.

List of Terms

AGL Above ground level
AIM Airman's Information Manual
ARTCC Air Route Traffic Control Center
ATC Air Traffic Control
ATIS Automatic Terminal Information Services
ATP Airline Transport Pilot
BCM Back Course Marker
CAB Civil Aeronautics Board
CAT Clear Air Turbulence
CG Center of Gravity
CHT Cylinder Head Temperature
DH Decision Height
DME Distance Measuring Equipment
Doppler An electronic device for long-range navigation
EFC Expect Further Clearance Time
EGT Exhaust Gas Temperature
EPR Engine Pressure Ratio
FAA Federal Aviation Administration
FAR Federal Air Regulation(s)
FL Altitude above sea level in hundreds of feet
Flight Safety International A company specializing in flight training
FOD Foreign Object Damage—litter or debris which can be ingested by a jet engine, causing damage
INS Inertial Navigation System—an electronic device for long-range navigation, particularly over water
KIAS Knots Indicated Airspeed
KTAS Knots True Airspeed
LOC Localizer

LIST OF TERMS

Mach   Speed in relation to the speed of sound
MDA   Minimum Descent Altitude
MAP   Missed Approach Point
MSA   Minimum Sector Altitude
NOS charts   National Oceanographic Service charts
NOTAMS   Notices to Airmen
NTSB   National Transportation Safety Board
OAT   Outside Air Temperature
Omega   An electronic device for long-range navigation, particularly over water
PPE   Pilot Proficiency Examiner
RVR   Runway Visual Range
SID   Standard Instrument Departure
STAR   Standard Terminal Arrival Route
VLF   Very Low Frequency
VMC   Minimum Control Speed
VTOL   Vertical Takeoff and Landing